SOUTH HOLLAND PUBLIC LIBRARY

3 1350 003

W9-BNN-342

SOUTH HOLLAND PUBLIC LIBRARY
708/331-5262
M-TH. 10-9, FRI. 10-6, SAT. 10-5
www.shlibrary.org

Cats Are Capable
of Mind Control

UberFacts @UberFacts · 1m

Cats Are Capable of Mind Control

And 1,000+ UberFacts You Never Knew You Needed to Know

KRIS SANCHEZ

   • • •

DEY ST.
AN IMPRINT OF WILLIAM MORROW *PUBLISHERS*

CATS ARE CAPABLE OF MIND CONTROL. Copyright © 2016 by UberFacts, LLC. All rights reserved. Printed in the United States of America. No part of this book may be used or reproduced in any manner whatsoever without written permission except in the case of brief quotations embodied in critical articles and reviews. For information address HarperCollins Publishers, 195 Broadway, New York, NY 10007.

HarperCollins books may be purchased for educational, business, or sales promotional use. For information please e-mail the Special Markets Department at SPsales@harpercollins.com.

FIRST EDITION   **3 1350 00359 6675**

Designed by Shannon Nicole Plunkett

Library of Congress Cataloging-in-Publication Data has been applied for.

ISBN 978-0-06-244116-4

16 17 18 19 20 OV/RRD   10 9 8 7 6 5 4 3 2 1

Contents

Introduction

UberFacts is the brainchild of social media addict/professional procrastinator Kris Sanchez—I originally started finding facts because I didn't want to study for a history exam my freshman year of college. Needless to say, I didn't pass the exam, never finished college, and this book almost didn't happen.

While searching for random facts about the number of insects on Earth or how honeybees reproduce isn't what you'd expect your average 18-year-old college student to do in his free time, it definitely paid off.

UberFacts now has more than 16 million followers across its social profiles and is dedicated to sharing the most bizarre, unimportant, and mind-boggling pieces of information anyone could ever ask for. The goal is to make learning fun, keep it light, and get readers genuinely interested in learning more about our weird world. The facts also make great icebreakers and can help you out on a bad date (believe me, I'm experienced when it comes to bad dates).

Of course, this intro wouldn't be complete without some shameless promotion, so make sure you find UberFacts on Twitter, Facebook, and Instagram!

Science

 Coffee has been found to reverse liver damage caused by alcohol.

A scientist has invented a bionic lens that can be surgically implanted into your eyeball so you would never have to wear glasses again.

 There are 3 trillion trees on the planet, but before humans started farming, there were twice as many.

Pigeons can be trained to detect breast cancer in mammogram scans.

The strongest material on Earth can be found in the teeth of sea mollusks called limpets. They can withstand the amount of pressure that turns carbon into diamonds. At only 1/100 the diameter of a human hair, their strength is equivalent to a piece of spaghetti holding up a hippopotamus.

This century, 14 animals have gone extinct, including a bright orange toad and 2 species of rhino.

The average male would have to drink 149 cans of Red Bull to die from a caffeine overdose. But you would definitely start vomiting before you could drink that much.

Male blue-capped cordon bleus (an African finch) tap dance to attract their mates.

The world's largest indoor farm is located in Miyagi Prefecture, Japan, and produces 10,000 heads of lettuce a day. Instead of solar power, the farm uses special LED lights that simulate night and day and emit an optimal wavelength of light for plant growth, allowing the farm to grow lettuce 2.5 times faster than an outdoor farm.

Male ducks have been observed engaging in necrophilic gay sex.

Vampire bats are the only bat species that can run.

The largest catfish on record is a Mekong giant catfish caught in Thailand in 2005. It weighed 646 pounds, which is about the same size as a grizzly bear, and was almost 9 feet long. Called a "royal fish," local fishermen sprinkled perfumed water on it as a sign of respect when it was caught.

Airplanes' black boxes are not actually black but orange.

Two engineers have died testing out the flying cars they invented.

The venomous bite of the bullet ant is considered to be the most painful sensation a human can experience. One scientist likened it to walking over coals with a 3-inch nail in your foot . . . for about 24 hours.

A glass ball bounces higher than a rubber ball, but a steel ball bounces even higher.

The world's largest underwater volcano, Tamu Massif, is the size of New Mexico.

The first animals in space were fruit flies, launched in a rocket by the United States in 1947. They were still living when recovered.

The only survivors of the 2003 *Columbia* space shuttle disaster were some parasitic worms called nematodes.

Perytons are radio signals that last only a few milliseconds and are caused by opening microwave doors prematurely. They're named after a so-called mythical creature most often attributed to Argentinian author Jorge Luis Borges.

A man spent 7 months constructing a 22-side Rubik's Cube. It exploded completely on the first turn.

It is possible to determine a person's gender from a fingerprint. Women secrete twice as many amino acids as men and in a distinct distribution.

Space smells like a combination of diesel fuel and barbecue, according to astronauts. The smell is caused by dying stars.

Hippos secrete a natural sunscreen in their sweat. The sweat contains microscopic structures that scatter sunlight, thereby preventing sunburns.

An island off the coast of Brazil known as "Snake Island" is said to have 1 snake per every square meter. Home to the extremely poisonous and also highly endangered golden lancehead viper, the island is completely off-limits to humans.

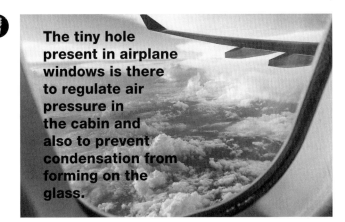

The tiny hole present in airplane windows is there to regulate air pressure in the cabin and also to prevent condensation from forming on the glass.

Some scientists believe that the earth is intent on self-destruction, to return to its original lifeless state. Called the Medea hypothesis, after the mythic Greek character who murdered her children, it's backed by the fact that there have been several mass extinctions triggered largely by microbes.

Physicists have concluded that the reason buttered toast lands buttered-side down is not because the butter makes it heavier, but because of Murphy's Law—everything that can go wrong will go wrong.

Contrary to popular belief, Kansas is not the flattest state in the United States. Florida is. Kansas is actually the seventh-flattest state.

There are 293 ways to make change for a dollar. But it's also possible to have coins that add up to more than a dollar and still not be able to make change.

While we think of space as a black vacuum, the average color of the universe is beige. Scientists call the color Cosmic Latte.

The 2011 Tōhoku earthquake in Japan was so strong that it shifted Earth's axis by about 6.5 inches and shortened the day by 1.8 microseconds.

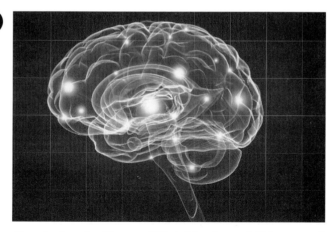

Google has built an artificial brain and it loves cat videos just as much as people do.

The Crooked Forest in Poland has over 400 pine trees bent at a 90° angle at the base. No one knows why, though it's been determined the bends occurred at some point in the 1930s, most likely man-made.

Scientists have discovered large quantities of water under the earth's surface. They're basically underground oceans.

Earth is 4.6 billion years old. If you scaled that down to 46 years, humans have been on the planet for only about 4 hours, and civilization as we know it has been only the last minute.

Lenticular clouds are rare formations that can look like UFOs. They are formed when wind encounters an obstruction such as a mountain or even a building and is disrupted into an eddy.

Astronomers have discovered a very peculiar light pattern around a star that is almost 1,500 light-years away. They spotted an unusual flickering pattern that, while possibly caused by comets, could be a sign of "alien superstructures" built around the star.

The Japanese island of Aogashima is a volcano within a volcano. In fact, it sits on the remains of 4 submarine volcanoes.

Biologists believe that replacing blood with younger blood may help reverse aging.

Most snakes have only 1 functioning lung. Their left lung is so small that in many species it doesn't serve a purpose.

Alligators have very, very small hearts. One hypothesis is that a small heart helps them hold their breaths while waiting for prey to come along.

Rapunzel, Rapunzel! A single strand of hair can hold up to 3 ounces of weight. That means the typical person's full head of hair can support up to 12 tons.

Each year, scientists discover an average of 15,000 new species.

Researchers study the respiratory systems of alligators by shotgunning fog machine juice into their mouths with an e-cigarette.

The average iceberg weighs 20 million tons.

The red-backed salamander doesn't have lungs and breathes through its skin.

 Baby Fitzroy River turtles breathe through their butts.

 Archaeologists in Wisconsin discovered 800-year-old seeds of an extinct squash and were able to grow the squash, bringing it out of extinction.

 The Omura's whale is about as long as a school bus, yet scientists could not locate a live whale until 2015. Prior to that, their existence was known only from dead specimens.

 The giant squid was long considered a mythological creature. There was never any hard proof they existed until one was caught on camera in 2013.

 The Aral Sea in Uzbekistan was once 1 of the 4 largest lakes in the world. But in the past decades it has shrunk to 10% of its original size after the Soviet Union decided to divert its rivers to grow rice in the desert.

A small, bucolic-looking stream in Yorkshire, England, has a tiny but extremely powerful chasm that leads to sudden death if you are unfortunate enough to fall in. It's said that no one who's ever fallen into the water has ever come out. Even their bodies are never found.

The center of the universe is located in Tulsa, Oklahoma. Well, not literally. That's the name of a curious phenomenon that was created by accident. Marked by a circle of bricks, sound is amplified and echoed within an 8-foot circle but is inaudible to anyone who stands outside the circle.

The symbol for plutonium in the periodic table is Pu because Glenn Seaborg, who discovered the element while working on the Manhattan Project, thought it would be funny to make the element "stinky." Pee-yew . . .

There is a black pebble beach on tropical island Koh Lipe, off the coast of Thailand. It is said that anyone who takes a stone off the island is cursed for life.

The Great Sand Sea desert in Egypt and Libya has many fragments of mysterious desert glass. It was likely formed 26 million years ago by meteorite impact, but geologists can't say for sure.

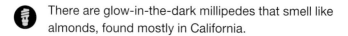 There are glow-in-the-dark millipedes that smell like almonds, found mostly in California.

At least 8 different species of tarantulas are bright blue.

It's possible to connect to the Internet in some way on 90% of Earth's land.

Mathematician Joseph Fourier believed that draping a blanket around him wherever he went would protect his health. Fourier died after he tripped and fell on the blanket.

Robert Fitzroy, captain of HMS *Beagle,* committed suicide after his expedition with Charles Darwin, partially because of his guilt over having helped Darwin develop his theory of evolution.

Cornell University scientists have created a functioning guitar the size of a human blood cell.

Pierre and Marie Curie's laboratory notebooks are still too radioactive to safely be handled without a protective suit.

In 2009, the Vatican announced that Charles Darwin's theories of evolution are compatible with Christian theology.

Jimmy Carter installed solar panels on the roof of the White House. Ronald Reagan had them removed.

Klerksdorp spheres are small round objects that have been found in pyrophyllite deposits about 3 billion years old. Nobody is exactly sure what they are, and while many mainstream geologists believe they are the product of natural processes in the earth, a large body of researchers cites them as evidence of prehuman intelligent life.

Staring at the sun can cause you to sneeze. It's due to the photic reflex, the mechanism of which is still largely unknown.

Sagittarius B2 is a molecular cloud made of ethanol alcohol near the center of the Milky Way. It is 3 times the mass of our sun.

Due to tidal acceleration, Earth's rotation has very gradually slowed over time. For example, 350 million years ago there would have been approximately 385 days per year.

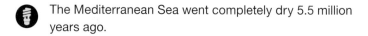 The Mediterranean Sea went completely dry 5.5 million years ago.

The sun's atmosphere is 600 times hotter than its surface.

The asteroid 2010 TK7 is 1,000 feet across and follows Earth in its orbit around the solar system.

Due to weather and topographical patterns that are not completely understood by scientists, there is almost constant lightning year-round at Maracaibo Lake in Venezuela.

NASA scientists believe free-floating planets, probably ejected from solar systems, may be more common in outer space than stars.

Many oranges are green when they're ripe. Scientists remove their chlorophyll to make them more appealing to North American consumers.

There is a cliff on one of Uranus's moons that's 12 times higher than the highest cliff in the Grand Canyon. It is believed to be the highest cliff in the solar system.

There were dragonflies the size of large birds 300 million years ago.

At 5,314 feet, Siberia's Lake Baikal is the deepest lake in the world.

The adult human body is made up of about 7 octillion atoms.

All the DNA in your body, if uncoiled, could stretch to Pluto and back 17 times.

The human body contains 10 times more bacterial cells than human cells.

At 1,600 miles in length, the Great Barrier Reef is the largest living structure in the world.

- There are 8 times as many atoms in a teaspoonful of water than there are teaspoons of water in the world's oceans.

- The average person walks the equivalent of 5 laps around the world during their lifetime.

- When helium gas is cooled to −452°F, it becomes a liquid that flows against gravity.

- Betelgeuse, the brightest star we can see, could become a supernova anytime between now and a few hundred thousand years from now. When it does, our sky will be continuously lit for 2 months.

- It takes a single blood cell 60 seconds to do a complete lap of the human circulatory system.

- While the universe seems infinite, there are only 100–200 galaxies that we know of.

- Game theory inventor Hugh Everett's multiple universes interpretation suggests that all possible alternative histories and realities are real, and is considered by most quantum physicists to be valid.

- Using a laptop that's connected to Wi-Fi decreases sperm mobility and fragments sperm DNA when placed on a male's lap.

The chills you get when listening to music are caused by your brain releasing dopamine, a neurotransmitter that causes pleasure.

 Women constitute 70% of Iranian university science and engineering students.

 A study showed that placing men's underarm sweat on a woman's lips helps regulate her menstrual cycle.

 Scientists have successfully used the AIDS virus to cure cancer in a mouse.

 In a real-life *Jurassic Park* scenario, Princeton scientists are attempting to re-create dinosaurs with chicken DNA.

 The Millennium Prize is a $1 million award given to whoever can solve any one of 7 math problems, but to date only 1 of the problems has been solved.

 Scientists have studied Ozzy Osbourne's DNA to determine how he survived so many years of extreme substance abuse.

A candle's flame contains millions of microscopic diamonds.

 The BBC conducted a study to see whether different foods would change the taste of semen. They did!

 A British research study found that watching a horror film prior to viewing abstract art enhances the enjoyment of the art for most people.

 ## The National Science Foundation awarded Indiana University $1 million to study suspicious memes and the viral spread of false ideas.

 Chinese scientist Tong Dizhou cloned the first fish, but the Maoist government forced him to abandon his research and become a janitor during the Cultural Revolution.

Russian scientist Alexander Bogdanov attempted to achieve immortality by transfusing his blood with that of young people. He died after accidentally transfusing his blood with that of a malaria victim.

After World War II, 1,500 German scientists were given amnesty on the condition that they work for the U.S. government as part of Operation Paperclip.

In 2015, the U.S. Department of Defense received $600 billion in funding. NASA received $17 billion.

Mexican scientists turned a shot of tequila into diamonds by heating it to 800°C.

In 2005, an Australian research institute published a study on the loss of teaspoons in the workplace.

Val Patterson, a scientist from Utah, confessed in his obituary that he never earned a Ph.D. nor graduated from college. He received his Ph.D. due to a clerical error but had completed only 3 years of college.

Researchers in Tokyo have developed a mirror that tweaks the viewer's reflection in real time to make it look like they're smiling.

When asked what his IQ was, Stephen Hawking said, "I have no idea. People who boast about their IQ are losers."

North Carolina has passed a law banning the use of scientific predictions as the basis for coastal policy.

A middle schooler's award-winning science project showed that, on average, 70% of the time the ice from fast-food restaurants is dirtier than the water from toilet bowls.

Only 1% of the world's water is drinkable. The rest is either polluted or salinated. And 99% of Earth's potable water remains in the polar ice caps.

There is more water in Earth's atmosphere than all the rivers on the planet's surface combined.

Alfred Nobel, of the Nobel Prize, lost his brother to an accident at his lab. They were working with liquid nitrogen and attempting to create a more stable explosive. Nobel was able to complete his research after the accident and invented dynamite.

Water is the only substance found on Earth's surface in all 3 forms—liquid, solid, and gas.

Gonorrhea bacteria can pull 100,000 times their own weight.

Hibernation is the practice of sleeping through the winter. Estivation is the practice of sleeping through the summer.

Russia has more surface area than Pluto.

Although Albert Einstein's research helped create the atomic bomb, he was not involved in the actual building of the bomb. The government believed his left-wing political views made him a security risk. Although he was a devoted pacifist, Einstein did write a letter to Franklin D. Roosevelt, urging him to prioritize the development of an atomic weapon. Einstein was deeply troubled by the knowledge that the Germans were also working on an atomic weapon and wanted to make sure the Americans beat them to it.

It takes 40,000 years for a photon to be spun out of the sun's core and make it to the surface. The rest of the journey to Earth's surface lasts about 8 minutes.

The average bolt of lightning contains enough energy to toast 100,000 pieces of bread.

Potatoes have 2 more chromosomes than humans do.

Human saliva contains opiorphin, a painkiller 6 times stronger than morphine.

At any given second roughly 100 lightning bolts are striking the earth.

Astronauts in space do not burp because there is no gravity to separate the gas from the liquid in their stomachs.

Human feces are 75% water. Bill Gates funded a machine that converts feces to water with the hope of solving the world's drinkable water crisis.

The red planet's seasons are much longer than ours. Due to Mars's elliptical orbit, spring and summer are longer in its northern hemisphere, while fall and winter are longer in the southern. They are also more intense, with hotter summers and colder winters.

Mars's surface is rich in iron, giving it a reddish color. This composition has led to cultures as far back as the Egyptians naming the planet after the color. The Egyptians named it Her Desher ("the red one"), the Chinese named it "fire star," and the Romans named it Mars, after their god of war (equivalent to Ares in Greek mythology).

A Martian day (24 hours and 40 minutes) is slightly longer than an Earth day (23 hours and 56 minutes).

The first spacecraft sent to explore Mars was the Soviet Union's Mars 1 in 1962. En route, mission control lost contact with the vehicle. The first American spacecraft to reach Mars was Mariner 4 in 1964, which sent back the first pictures of the red planet.

Phobos, the larger of the 2 Martian moons, orbits the planet so quickly, it would set twice (in the east) and rise once (in the west) every day. Deimos, the smaller moon, is so small that an astronaut on Mars would see it as a full moon just about as brightly as we see Venus in our night sky. Scientists are unsure whether both these moons are captured asteroids or true moons. Each orbit, Phobos is drawn closer to Mars by the planet's gravity. (It spirals inward 6 feet [1.8 meters] every 100 years.) Many millennia in the future, Phobos will likely crash into Mars or break up above the planet's surface, creating a ring. Could spell trouble for future Martian colonists!

Lower gravity on Mars means you would be able to hop around much easier than on Earth. Its 62.5% less gravity means 100 pounds on Earth are equivalent to about 38 pounds on Mars.

Besides Earth, Mars is the only planet in the solar system to have polar ice caps.

Uranus is not visible to the naked eye and was the first planet discovered with the use of a telescope. It was officially discovered in 1781 by Sir William Herschel, who also went on to discover the moons of Saturn.

Uranus is the smallest of the 4 "giants" (the others being Jupiter, Saturn, and Neptune), but it is still several times larger than Earth. It has a diameter of 29,297 miles (47,150 kilometers), compared with Earth's diameter of just under 8,000 miles (12,760 kilometers).

Uranus has a total of 27 moons, most of which are named after characters in Shakespeare's *A Midsummer Night's Dream*. The 5 major moons are called Titania, Oberon, Miranda, Ariel, and Umbriel. Umbriel is not from Shakespeare but is the "melancholy sprite" in a poem by Alexander Pope.

Only 1 spacecraft has flown by Uranus. In 1986, Voyager 2 swept past the planet at a distance of 81,500 kilometers. It returned the first close-up images of the planet, its moons, and its rings.

Uranus makes 1 trip around the sun every 84 Earth years. During some periods in its orbit, one of its poles points directly at the sun and gets about 42 years of direct sunlight. The rest of the time it is in darkness.

Arabic numerals, like the ones we use today in English, were actually invented by mathematicians in India.

Earth spins at around 1,000 mph and hurtles through space in its orbit around the sun at about 67,000 mph.

The solar system is 4.56 billion years old.

The tallest tree ever measured was an Australian eucalyptus tree. It was spotted by forester William Ferguson in 1872 and towered at 435 feet tall.

Methuselah, a bristlecone pine, is thought to be the world's oldest tree and is estimated to be just shy of 5,000 years old. Methuselah resides in California's White Mountains, but the tree's exact location is kept secret for its protection.

A raindrop falls at a maximum speed of 18 mph.

Albert Einstein was a great scientist but, by all accounts, a terrible husband. He cheated on his wife, belittled her scientific achievements, and refused to help out around the house. He also made her sign a contract in which she would agree to leave the room or stop talking if he told her to.

All humans are radioactive. Sleeping in the same bed with an average-size adult exposes you to about 1 millirem of radiation.

Scientists have been able to genetically modify mosquitos so that they are unable to transmit malaria to humans.

The average temperatures for the year 2015 were so hot that the United Nations declared it the warmest year on record before November had even ended.

Ants can link their bodies together to form bridges. Scientists have been studying this phenomenon, hoping it will lead to developments in robotics.

There is an enormous (below 220 Dobson units) hole in the Antarctic ozone layer, but scientists predict it will recover by the year 2040, if at all.

Looks like we're one step closer to Terminators. Robots have learned to say no to commands from humans.

A jiffy is a real measure of time. In quantum physics, a jiffy represents the time it takes for light to travel 1 Fermi (about the size of a nucleon, or 10^{-15} meters). So a jiffy is 3×10^{-24} seconds, but is currently impossible to measure with today's technology.

In the 1990s, a biotech company developed genetically modified bacteria to get rid of crop residue after harvest. However, it was never tested outside of the laboratory out of the fear that it could possibly destroy all the plant matter in the world.

The square root of –1 is i.

The first webcam was used to broadcast the coffeepot in the Trojan Room lab at Cambridge University. It was set up so people could tell if the coffeepot was full.

Physicists have figured out how to levitate frogs with magnets.

Earth's magnetic field has decreased by 5% over the past 100 years.

The computers in today's cars are more powerful than the computer system that guided the Apollo spacecraft to the moon.

The air pressure weighing down on you at any given time is the equivalent of a small car, about 1 ton.

The longest-running experiment is the pitch drop experiment. A funnel holding a sample of tar pitch sealed in a glass jar shows how some substances that appear solid are actually liquid. It takes 9 years for a single drop to form. The last drop happened in 2014.

Corpse flowers bloom only once every 7–10 years. They get their name from their distinct smell, which is very similar to the smell of rotting meat.

 Slime molds, a simple organism that consists of an acellular mass of creeping, jellylike protoplasm containing nuclei, can solve puzzles.

 The tallest mountain in the solar system, Olympus Mons, is a shield volcano, similar to those in Hawaii and many Pacific islands. It stands at 13.2 miles (21 kilometers) above sea level on the Martian surface. Three times taller than Mount Everest, Olympus Mons's surface area is the same as the entire state of Arizona.

 The world's oldest clove tree is nicknamed Afo. Found on the Indonesian island Ternate, it is between 350 and 400 years old.

The insect *Issus coleoptratus* has joints that function like mechanical gears.

Dung beetles use the Milky Way to navigate.

The first-ever recorded use of the word "scientist" was in the 1883 essay "A Plea for Pure Science" by H. A. Rowland of Baltimore, Maryland.

In 2012, 10-year-old Kansas City elementary school student Clara Lazen accidentally created a new molecule in science class, tetranitratoxycarbon, using a ball-and-stick model.

In 2009, 41 new mammals were discovered.

Toxineering is a new branch of science dedicated to transforming animal venoms into painkilling medicines.

Researchers at the Bristol Robotics Laboratory have developed a new method of charging cell phones with human urine.

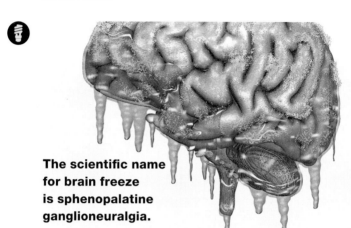

The scientific name for brain freeze is sphenopalatine ganglioneuralgia.

Geophysicists in Germany have successfully used the carbon in peanut butter to manufacture diamonds.

A bolt of lightning is 5 times hotter than the surface of the sun.

Because of its high protein and fatty acid content, eating salmon helps hair grow faster and stronger.

Researchers from Heidelberg University Hospital have determined that it takes 6 minutes for alcohol to impact human brain cells.

Rainwater contains high quantities of vitamin B_{12}.

Millions of sunflowers have been planted around the Fukushima nuclear power plant to help clean up radioactive waste. The plants are known to soak up toxins.

Woodpeckers slam their beaks into trees 8,000–12,000 times per day at 1,000 times the force of gravity.

Sniffer bees and wasps have been trained to detect illegal drugs and explosive materials, and their ability to do so rivals sniffer dogs.

The tomato has more genes than a human being.

Challenger Deep, the deepest part of the Mariana Trench, is so deep that you could fit 29 Empire State Buildings, top to bottom, inside of it before they reached the ocean's surface.

The existence of Neptune was predicted mathematically in the 17th century, but was not observed with a telescope until 1846.

Scientists have identified approximately 173,000 venomous species in the world.

The sun is believed to have completed between 18 and 20 orbits around the Milky Way in its lifetime, and about 1/250 of an orbit since the emergence of the human race.

Gasoline has a complex chemical structure. It typically contains 150 carbon compounds, but can contain up to 1,000.

The smell of rain is produced by soil-dwelling bacteria called actinomycetes.

Prior to his career as "the science guy," Bill Nye was a stand-up comedian.

In 1992, 29,000 rubber ducks were lost at sea, and they are still being discovered in unexpected places.

When you donate your body to science, it can be used as a crash test dummy.

Science fiction writer Isaac Asimov died in 1992 from complications related to HIV, which he acquired through a blood transfusion.

Ancient penguins grew to 5 feet in height.

The latest fad in Silicon Valley performance enhancements is microdosing with LSD and psychedelic mushrooms. Users claim it makes them more productive, more creative, and generally better at everything.

Alligators either can't stick out their tongues or don't even have them. Scientists are unsure.

Peanut oil can be used to make nitroglycerin, an essential ingredient in dynamite.

Venus is the only planet in the solar system that rotates clockwise.

Over half of the American coastline is in Alaska.

St. Louis's Gateway Arch looks taller than it is wide, but that's an optical illusion. It is 630 feet tall and 630 feet wide.

Caño Cristales in Colombia is said to be the most beautiful river in the world. The water is exceedingly clean, and at certain points in the year the riverbed turns bright red, blue, and yellow.

Before 2009, Oklahoma had about 2 earthquakes a year. In the last 2 years there have been an average of 2 a day. It's believed they are caused by fracking.

In 1758, 274 moths were given scientific names.

The average European is 2.7% Neanderthal.

There are 50,000 earthquakes on Earth each year.

It would take 7 billion particles of fog to fill a teaspoon.

Globally, only 2% of the population has green eyes.

The mass of Earth is 6,000,000,000,000,000,000,000,000 kilograms.

1. **According to most scientific estimates, how old is Earth?**

a. 1.56 billion years old   c. 3.56 billion years old

b. 2.56 billion years old   d. 4.56 billion years old

2. **What is the only planet in our solar system—other than Earth—to have polar ice caps?**

a. Venus   c. Mars

b. Saturn   d. Pluto

3. **Hibernation is the practice of sleeping through the winter. What is the practice of sleeping through the summer?**

a. Estivation   c. Helovation

b. Reverse hibernation   d. Photosynthesis

4. **Who, when asked about their exact IQ, said, "I have no idea. People who boast about their IQ are losers"?**

a. Stephen Hawking   c. Albert Einstein

b. Bill Gates   d. Mark Zuckerberg

5. **Airplanes' flight recorders are what color?**

a. Black   c. Orange

b. Blue   d. Pink

6. **How many ways are there to make change for a dollar?**

a. 193   c. 493

b. 293   d. 1093

7. **In 2009, the Vatican announced that this man's theories are in fact compatible with Christian theology. Who is this person?**

a. Albert Einstein

b. Charles Darwin

c. Jonas Salk

d. The Dalai Lama

8. **At 5,314 feet, what is the deepest lake in the world?**

a. Lake Michigan

b. Lake Titicaca

c. Lake Baikal

d. Lake Champlain

History

Arguably the friendliest territorial dispute is over an uninhabited island called Hans Island, claimed by both Denmark and Canada. When Danish military go there, they leave a bottle of schnapps, and when the Canadians go there, they leave a bottle of Canadian Club whisky and a sign saying WELCOME TO CANADA.

In colonial India, the British government tried to control the poisonous cobra population by offering a bounty for dead cobras. Indians realized they could collect the reward simply by breeding cobras and killing them. When the government realized the scheme, it stopped offering the bounty. Breeders then released all their cobras into the wild, and the cobra population was higher than it had been before. This is the origin of what's called the cobra effect, or measures taken that end up making the problem worse.

French apothecary and reputed seer Nostradamus successfully predicted his own death. He died a day after making the prediction.

There's no proof the Pilgrims actually landed on Plymouth Rock. The Thanksgiving tradition dates to the mid-18th century when Puritan patricians would take their children to the rock every Thanksgiving and lecture them about history.

During World War II, most American cars had wooden bumpers. Car owners turned in their chrome bumpers to aid in the war effort.

Juvenile offender Willie Francis had the bad luck of being executed twice. His first time in the electric chair failed because the executioner was drunk and set it up improperly. He then brought his case to the Supreme Court, arguing a second execution would constitute double jeopardy. He lost and was successfully executed by electric chair in 1947.

Great Britain briefly had a Cones Hotline in the early 1990s. It was a special number citizens could call if they saw traffic cones on the road for no reason. It was disbanded after 3 years because almost no one ever called it.

The Egyptian pharaoh Ramses I, the founder of one of Egypt's most powerful dynasties, was not born a prince. He was adopted as an adult by the heirless military ruler Horemheb, who became pharaoh after the teenage King Tut died an untimely death.

In the 16th century, Hans Steininger was famous for having one of the world's longest beards. It was 4.5 feet long and kept rolled up in a pouch. When a fire broke out in his town, Hans tripped over his own beard, broke his neck, and died.

During the Great Depression in London, men who couldn't afford to sleep in a bed at cheap lodging establishments could pay twopence for a spot on a bench with a clothesline tied in front of them, so they could sleep while hanging over the rope.

In 1508, the French village of Autun tried to summon its rats to court on the charge of eating the farmers' barley crop. When the rats failed to appear in court, their lawyer, Barthélemy de Chasseneuz, convinced the judge that each rat had to be served a summons individually. Of course, the rats failed to appear a second time. Their lawyer argued they couldn't make it because they feared for their safety from the local cats.

A New Jersey man flunked out of law school and subsequently sued the school for having accepted him in the first place.

In 1958, the U.S. Air Force accidentally dropped a nuclear bomb on a house in South Carolina. A 7,600-pound atomic warhead fell out of a plane during a test run. Fortunately, the bomb was unarmed and the only casualties were a few chickens.

Until the beginning of the 19th century, America's largest Jewish population lived in Charleston, South Carolina.

There is no proof pirates ever said "Arrrrr" or "Shiver me timbers."

 The presidential pardon of a turkey on Thanksgiving only became tradition with the 41st president, George H. W. Bush, who pardoned a turkey every year of his presidency.

 An Irishman named Tom Johnson was offered the modern equivalent of $3 million to rescue the deposed Napoleon from his exile on the remote and heavily guarded island St. Helena. He proposed to do it by submarine, which he claimed to have built himself, decades before functional submarines existed.

 In Victorian London, people were paid to collect dog poop. It was used to tan leather. Surprisingly, it was a sought-after job, since it paid fairly well.

Dean Kamen, the inventor of the Segway, owns an island off the coast of Connecticut called North Dumpling Island. The island has its own constitution, flag, currency, and even navy (though the navy is made up of just 1 boat). It also has a replica of Stonehenge. Kamen refers to himself as Lord Dumpling, though the island is not technically recognized as separate from the United States.

The most remote island on the planet is a 19-square-mile hunk of rock and ice in the South Atlantic, 1,000 miles from Antarctica and 1,400 miles from South Africa. Yet when South African officials studied the island in 1964, they found an abandoned rowboat and a set of oars. Despite evidence that people had been in the boat at the time of its landing, no human remains were ever found and no one knows how the boat got there.

Bras were called "breast bags" in medieval times.

The longest prison sentence served in the United States was 68 years and 245 days by Paul Geidel. Convicted of murder at 17 years old in 1911, Geidel was released when he was 86, even though he had been offered parole 6 years earlier and declined.

John Quincy Adams was the first U.S. president to appear publicly in pants instead of breeches.

The largest medieval city in present-day America was called Cahokia, located not far from where St. Louis is now. At its height, circa 1100–1200, it had a population larger than London's. But a century later the city was largely abandoned.

It's generally believed that the ancient Egyptians were the first to domesticate cats, 4,000 years ago. But archaeological research has found that farmers in China kept cats as pets 5,300 years ago.

The Chicago River was once flooded with onions. An onion farmer turned commodities trader in the 1930s was able to manipulate the onion market to the point where bags of onions sold for less than the bags themselves cost, and Chicago was completely overrun with onions. To this day, it is illegal to trade in onion futures.

The original name of Liberty Island, home of the Statue of Liberty, was Bedloe's Island.

The name Jolly Roger, referring to the skull and crossbones pirate flag, is thought to have come from an old nickname for the devil, Old Roger. Earlier pirate flag designs showed the devil or a skeleton thrusting a spear through a heart.

In the 1970s, geologists discovered a Russian family in the Siberian taiga 150 miles away from the nearest human settlement. Religious fundamentalists, they had fled civilization after the Communist revolution and were completely unaware of World War II, having been totally isolated for 40 years.

Jimmy Carter reported seeing a UFO before he was president, in 1969. He later said he did not believe it was an alien spacecraft.

The first cross-country highway was not built by the government, but by the auto industry. Named the Lincoln Highway, the first official transcontinental road trip took place in 1915—it spanned 3,384 miles and took 102 days to complete.

Trains on the fastest railway in Europe can reach speeds up to 250 mph. Meanwhile, the fastest train in the United States averages only 80 mph.

The Waldorf Astoria hotel once had its own private railroad track at Grand Central so that its guests could clandestinely enter and exit New York City. Largely abandoned now, it operates only when the president is in town, in case the need arises for an emergency exit.

Franklin D. Roosevelt moved Thanksgiving up a week in 1939 as an economic stimulus. While Black Friday wasn't quite the tradition back then that it is today, it still marked the beginning of the holiday shopping season and Roosevelt thought it would encourage more spending. Congress moved it back 2 years later.

Moscow has a second, possibly abandoned but definitely classified, metro system that was built for the sole use of Communist officials. The Russian government will neither confirm nor deny its existence.

In 1996, archaeologists unearthed a 7,000-year-old mass gravesite in Germany that showed evidence of cannibalism, with the remains of at least 1,000 people, from small children to the elderly, who were butchered and eaten, most likely in a ritualistic manner.

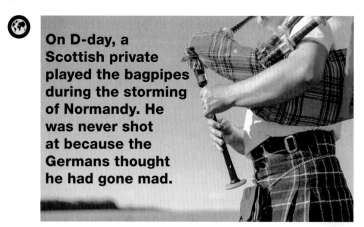

On D-day, a Scottish private played the bagpipes during the storming of Normandy. He was never shot at because the Germans thought he had gone mad.

The fork was invented in ancient China.

Henry Ford tried to build an American industrialist utopia in the Brazilian jungle, naming it Fordlandia. The local workers ended up rebelling because Ford had outlawed drinking, smoking, and women.

The amputated arm of Civil War general Stonewall Jackson is buried separately from his body.

New York City's oldest house is a cottage in Queens near LaGuardia Airport. It was built in 1654 by the Rikers family (the same family that gave Rikers Island its name) and even has a family cemetery in the backyard.

Nüshu is a language created by women in 15th-century China, and it was used in secret for centuries among women to educate themselves without men finding out about it.

Currency shaped like knives was in circulation in China from 600 to 200 B.C.

In 1533, Shaolin monks were called in to fight off Japanese pirates attacking the Chinese coast.

During the 1909 World's Fair in Seattle, a 1-month-old baby was given away in a raffle, and historians don't know what happened to the child.

The first slaves in the Americas were Irish.

The first recorded mooning happened in A.D. 66, when a Roman soldier mooned a group of Jewish pilgrims.

When Julius Caesar was kidnapped by pirates, he told them he was worth twice as much as their asking ransom.

In the Middle Ages, people stored their coats in the toilet room, because the ammonia from urine would kill the fleas.

In ancient Greece, wine was diluted with 3 times more water than wine.

King Tut had his own personal nose-picker.

During the early 20th century, duels with wax bullets were a popular recreational activity.

Spartan women would shave their heads on their wedding day in order to help Spartan men transition from homosexual to heterosexual intercourse.

When smoked raw, tobacco has psychedelic properties. This is how Native Americans smoked it during ceremonies.

In 1857, a group of Mormons and Native Americans attacked a group of settlers in Utah's Great Basin, killing between 120 and 140 men, women, and children.

The Roman emperor Elagabalus often dressed in drag and promised physicians gold if they could perform a sex change on him.

Four Auschwitz inmates successfully escaped by stealing SS officer uniforms and driving a stolen Nazi car through the camp's front gate.

The Allies parachuted dummies over Normandy Beach on D-day to distract Nazi gunners from the real paratroopers.

The eruption of Mt. Tambora led to the unseasonably cool summer of 1816. The inclement weather led to a cholera outbreak, which forced Lord Byron and the Shelleys out of London, and all three went on to produce legendary literary works. The eruption also caused London's sunsets to be particularly colorful, inspiring J. M. W. Turner's sunset paintings.

- Cruentation was the medieval belief that the body of a murder victim would spontaneously bleed in the presence of the murderer.

- The Roman emperor Claudius's romantic life was considered atypical because he only had sexual relationships with women.

- The British army recruited 90,000 Chinese laborers to dig trenches during World War I.

- The United States paid Iran $61.8 million in 1988 in reparation for shooting down a civilian plane, but it did not issue an apology.

- The earliest surviving written music is a Hurrian hymn from 1300 B.C.

- The first downloadable content was available in 1982.

- Eight people drowned in the London beer flood of 1814, when all the vats in a brewery ruptured and flowed down a Tottenham street.

- In the 1950s and 1960s, the Canadian government used a screening process called the "fruit machine" to identify homosexual men and bar them from public service.

- Inhabitants of the Chinese town of Liqian have blond hair and white skin. Historians have speculated they are the descendants of a lost Roman settlement.

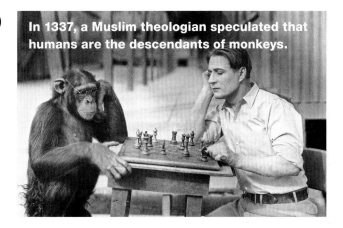

In 1337, a Muslim theologian speculated that humans are the descendants of monkeys.

The top speed of the first American car race in 1895 was 7 mph.

The Tollund Man was a 4th-century B.C. man who was naturally mummified. He was mummified so well, in fact, that the modern people who accidentally discovered him thought he was a recent murder victim.

Ancient Greek statues were originally painted with bright colors.

Ötzi, the oldest intact mummy ever discovered, had tattoos.

Until 1916, soldiers in the British army were required to sport mustaches.

Germany made its final reparation for World War I, under the Treaty of Versailles, in 2010.

- Muslims and Jews fought side by side against Christians during the Crusades.

- In the 19th century, doctors treated hysteria in women by inducing orgasms. This is how the vibrator was invented.

- Punishment for a female serial killer in ancient Rome was rape by a giraffe.

- Ancient Mesoamericans carved mirrors from shiny stones and considered them portals to other dimensions.

- The 15-year-old Roman emperor Heliogabalus was a practical joker of epic proportions. In addition to creating the whoopee cushion, he forced dinner guests into a room containing toothless leopards. He was assassinated at age 18.

- In 1252, the king of Norway gave Henry III a polar bear as a gift. The bear lived in the Tower of London, attached to a very long leash so he could catch fish from the Thames.

- The word "ostracism" derives from a classical democratic process, in which Athenians voted once a year to expel the most annoying citizen from the city-state.

In the 4th century B.C., Greek arsonist Herostratus burned down the Temple of Artemis, one of the 7 wonders of the world. After he was executed, a law was passed making it illegal to say the arsonist's name.

Mithridates VI of Pontus was terrified of being poisoned, so he took small doses of various poisons throughout his life to develop immunity. When the Romans captured him, he tried to commit suicide by poison, but survived due to his immunity.

Only 2 groups of Spartans were entitled to have their names inscribed on their tombstones: women who died in childbirth and soldiers killed in battle.

As he walked on British soil for the first time, William the Conqueror slipped and fell down. To play it off, he grabbed 2 fistfuls of dirt, raised his arms, and declared, "England is ours!"

In 324 B.C., Alexander the Great held a drinking contest in honor of the deceased Indian Brahman Calanus; 42 of his soldiers drank themselves to death.

Purple was the royal color because of its rarity. Prior to the creation of synthetic dyes, purple could only be extracted from sea snails—a most laborious task.

In 1982, Key West attempted to secede from the United States.

Mail was delivered 12 times per day in Victorian London.

In ancient Rome, togas were washed with ammonia from human urine—the origin of modern dry cleaning. The urine collectors were taxed.

During the black death, incoming ships were forced to wait for 40 days to prevent possible infection. The Italian word for 40, *quaranta,* is the origin of the word "quarantine."

The first African American to win the Nobel Peace Prize was Ralph Bunche, who won in 1950 for his mediation work in Israel. He was also involved in the formation of the United Nations.

Windsor Castle is the longest-occupied castle in Europe, having had residents since the 11th century. The royal family slept there during World War II, but this fact was kept secret in the hopes of avoiding aerial bombardment by the Nazis.

On May 5, 1868, 3 years after the Civil War ended, the head of an organization of Union veterans—the Grand Army of the Republic—established Decoration Day as a time for the nation to decorate the graves of the deceased with flowers. Confederate states also created their own holidays to memorialize those who had perished, but by the 20th century those holidays were combined into Memorial Day.

Benito Mussolini once said, "Anti-Semitism does not exist in Italy . . . Italians of Jewish birth have shown themselves good citizens, and they fought bravely in the war. Many of them occupy leading positions in the universities, in the army, in the banks." He later changed his views to curry favor with Adolf Hitler and Nazi Germany.

Turkmenian president Saparmurat "Turkmenbashi" Niyazov (r. 1985–2006) banned car radios, lip-synching, video games, opera, and long beards. He also renamed the month of January after himself on the Turkmenian calendar. In 2005, he launched a copy of his book into space. The state-owned newspaper, *Neitralny Turkmenistan,* reported, "The book that conquered the hearts of millions on Earth is now conquering space."

Haitian dictator François "Papa Doc" Duvalier changed the Lord's Prayer and made it about himself. The version he had children recite in school started out: "Our Doc, who art in the National Palace for life, hallowed be Thy name by present and future generations. Thy will be done in Port-au-Prince as it is in the provinces. Give us this day our new Haiti and forgive not the trespasses of those anti-patriots who daily spit upon our country . . ."

Before becoming a brutal Syrian dictator, Bashar al-Assad was a fully credentialed ophthalmologist working at a clinic in London. It was only after his brother, the successor to his father's rule, died in a car accident that he was pressured into taking on the role.

Romanian dictator Nicolae Ceaușescu was so paranoid about people poisoning his clothing, he had his suits protected by armed guards and never wore the same suit twice.

Joseph Stalin was born Iosif Vissarionovich Dzhugashvili. He changed his name to Stalin in his 30s. *Stalin* is Russian for "steel-man" or "man of steel," so Stalin—inadvertently or purposefully—ripped off Superman.

Canada had a brief flirtation with Nazism in the 1930s that is mainly attributed to Adrien Arcand, or "Canada's Fuhrer." Prime Minister William Lyon MacKenzie King expressed a positive response upon meeting Adolf Hitler in the late 1930s. Arcand, then a Montreal journalist, declared himself head of Canada's "National Unity" Fascist party. His rallies drew thousands of people until the war broke out and he was thrown in jail. After his release he remained committed to fascism until his death in 1967.

President Harry Truman's nickname for Stalin was "the little squirt." The name was well deserved: Stalin was only 5-foot-4.

Stalin is attributed the quote: "One death is a tragedy; one million is a statistic." It's believed that anywhere between 3 and 60 million Russians died during Stalin's reign, from famine, execution, and being worked to death in the gulags.

Abraham Lincoln was the first president to pardon a turkey. In 1863, when a live turkey was brought home for the Christmas dinner, Lincoln's son Tad pleaded for the turkey to be spared, and Lincoln obliged.

George Washington was the only president who did not live in the White House, though he did choose the site for the historic building and approve the plans for its design.

John Adams was the first president to reside in the White House, having moved in November 1, 1800.

 The modern 50-star American flag we know and use today was designed by 17-year-old high school student Robert G. Heft in 1958. Heft originally created the flag as a school project. He got a B– on the project, but the teacher agreed to change his grade after Congress adopted his flag design. Although Heft is usually given credit for the flag design, several identical designs were submitted at around the same time. The flag was being reworked because Hawaii and Alaska were joining the union.

The White House didn't get its first coat of white paint until 1818. Before that the lye used to weatherproof the stone facade gave the White House its distinct coloration.

The White House was built by slaves, free blacks, local artisans, and newly arrived European immigrants seeking work. Construction began in 1792 and ended in 1800.

Theodore Roosevelt was the first president to have a round-the-clock Secret Service detail. He described his security detail as a "very small but very necessary thorn in the flesh" in a 1906 letter.

During World War II, the U.S. government experimented with, but ultimately decided against, attaching napalm bombs to bats and unleashing them over Japan.

Mexican general Antonio López de Santa Anna had his leg amputated after he was hit with cannon fire. He gave the leg a formal burial with full military honors.

New Mexico State University was supposed to issue a diploma to its first graduate in 1863, the sole member of his graduating class, but he was killed in a gunfight before the ceremony.

Saddam Hussein was given a key to the city of Detroit in 1980.

Electricity was installed in the White House in 1891. President Benjamin Harrison and his wife, Caroline, were scared to use the light switches for fear of getting shocked.

It is widely believed that ancient Persians debated important issues twice, once drunk and once sober. Only ideas that remained credible in both states were deemed valid.

The world's longest war was the Three Hundred and Thirty-Five Years' War fought between Sicily and the Netherlands. It ended with a peace treaty in 1986. It was a completely bloodless conflict with 0 casualties.

The world's shortest war lasted 38 minutes. It was the Anglo-Zanzibar War of 1896.

Albert Einstein was offered the role of Israel's second president in 1952. He said, "My relationship to the Jewish people has become my strongest human bond, ever since I became fully aware of our precarious situation among the nations of the world," but ultimately declined the offer, believing he lacked the interpersonal skills to lead.

The U.S. Army's code name for the building of the first atomic weapon was the Manhattan Project. The first 2 bombs were code-named Fat Man and Little Boy.

Karl Marx was a European news correspondent for the *New-York Daily Tribune* in the 1850s.

Many in the U.S. government believed Albert Einstein to be a Soviet spy. His phone was tapped, his mail was read, and even his nephew's home was bugged. This was due mainly to his controversial political views that included universal civil rights and nuclear disarmament. The FBI dossier on Einstein goes so far as to claim he is more radical "than Stalin himself." No proof of Einstein as a spy has ever surfaced.

Kim Jong Il wrote 6 operas. He also published a book on the art form, entitled *On the Art of Opera: Talk to Creative Workers in the Field of Art and Literature, September 4–6, 1974.*

In 1978, Kim Jong Il kidnapped 2 famous South Korean filmmakers and forced them to work on North Korean propaganda films. The harrowing ordeal included being forced to remarry (they had previously married and divorced) and working on Kim's personal film projects. They escaped by fleeing to the American embassy in Vienna in 1986. They had gone to Vienna on Kim's behalf to promote a North Korean film he had forced them to make.

Kim Jong Il was scared to fly and built himself a special armored train car. He died, of natural causes, in that very train car in December 2011.

The B-29 bomber that dropped the first atomic bomb on Hiroshima was nicknamed the *Enola Gay*. The bomber that dropped the Nagasaki bomb was nicknamed the *Bockscar*.

When the American Civil War started, Confederate Robert E. Lee owned no slaves. Union general Ulysses S. Grant did. Grant freed his slaves only after the passage of the 13th Amendment, which banned slavery. When asked why he hadn't freed his slaves sooner, Grant said, "Good help is so hard to come by these days." Lee said, "Slavery as an institution, is a moral & political evil in any Country," in a letter to his wife dated December 27, 1856.

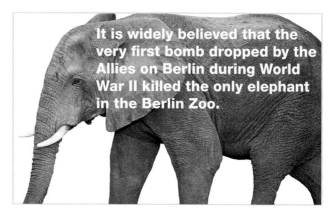

It is widely believed that the very first bomb dropped by the Allies on Berlin during World War II killed the only elephant in the Berlin Zoo.

During World War II, American censors attempted to ban all things "German." German measles were renamed "liberty measles," sauerkraut was renamed "liberty cabbage," and dachshunds were renamed "liberty hounds," much like the ballyhooed "Freedom fries," renamed to protest France's lack of support for the Iraq War.

The parliament of Iceland—the Althing—is the oldest active parliament in the world. It was established in 930.

Archaeologists have excavated the ruins of a military stronghold that defended the Temple Mount in Jerusalem. The site was found underneath a parking lot.

ISIS's biggest enemy might be a Marxist feminist collective of Kurdish rebels who live in Syria. The Women's Protection Unit, or YPJ, is an all-volunteer army of women who fight alongside their male counterparts. Many ISIS fighters believe that if they are killed by a woman, they will not go to Heaven, making the YPJ a particularly terrifying fighting force.

During renovations at the University of Virginia, construction workers found a hidden chemical lab dating back to the 1840s.

Scuba divers recently discovered a 10,000-year-old limestone monolith, similar to those at Stonehenge, at the bottom of the Mediterranean Sea. It is believed to have been carved by the Mesolithic people who lived on islands between Tunisia and Sicily that are now submerged.

Archaeologists found an old butcher shop in Greece that dates back 500,000 years. There is evidence that elephants were slaughtered and carved by early hominids.

The first conversations ever were most likely about rocks. Specifically, turning rocks into tools.

 The oldest time-telling device is a kind of sundial found in Egypt, thought to date back to 1500 B.C. It's a bent T shape that had to be turned around after noon.

 The remains of England's King Richard III were found buried under a parking lot in Leicester in 2013. The king met his death on the battlefield in 1485.

 The Acting Witan of Mercia is an English political organization that claims the British government illegally occupies the region of Mercia. As such, it has its own constitution and currency.

 The Great Molasses Flood of 1919, in which 21 people died, occurred in Boston after a tank holding as much as 2.3 million gallons of molasses collapsed.

Barbed wire was invented in 1845 and was largely responsible for putting cowboys out of business since it provided cheap and easy fencing.

Horace Fletcher was a Victorian nutritionist nicknamed the "Great Masticator." He advocated for chewing each mouthful of food 100 times.

Moulay Ismaïl, who ruled Morocco in the 17th century, is said to have fathered 867 children: 525 sons and 342 daughters.

Leonard Jones ran for president several times in the 19th century. His sole platform was that he would live forever. He died of pneumonia in 1868 after refusing medical treatment.

An unfortunate job in the 1930s was testing safety glass—glass that cracks but does not shatter. The glass was tested by smashing a man's head into it.

A statistician calculated that the odds that Mikhail Gorbachev is the Antichrist are 710,609,175,188,282,000 to 1.

The Dundee-Happy Hollow Historic District is a neighborhood in Omaha, Nebraska, that was firebombed by the Japanese during World War II. No one was hurt, and the media never reported it to prevent a panic.

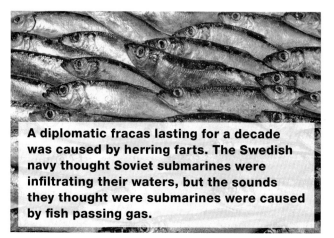

A diplomatic fracas lasting for a decade was caused by herring farts. The Swedish navy thought Soviet submarines were infiltrating their waters, but the sounds they thought were submarines were caused by fish passing gas.

The Italian government includes revenue from prostitution, smuggling, and illegal drug sales in its GDP accounting.

Police in Belarus arrested a one-armed man for allegedly clapping in public at a rally to protest President Alexander Lukashenko's policies.

The U.S. Government Accountability Office released a report in 2012 titled *Actions Needed to Evaluate the Impact of Efforts to Estimate Costs of Reports and Studies*. In other words, a report about a report about reports.

The first U.S. town to be completely lit by electric streetlights was Wabash, Indiana, in 1880. It had a population of 320 at the time.

When the Milton Bradley Company unveiled Twister in 1966—the first game to use human pieces—it was derided as "sex in a box."

The Great Fire of London in 1666 destroyed 13,000 houses and left 70,000 people homeless, but killed only 8 people.

The first true concept of 0 was developed by ancient mathematicians in what is now India, at some point between the 5th and 7th centuries.

At the start of World War I, the U.S. Air Force had 12 planes and 18 pilots to fly them.

King George I of Great Britain was born in Hanover, Germany.

Lord Byron attended Trinity College, where he kept a pet bear in his dorm room.

President Ronald Reagan worked as a lifeguard for 7 summers as a teenager, and during this time he was credited with saving 77 lives.

In 1929, 2 Princeton researchers wired a cat to the phone lines to turn the animal into a working phone.

Although the episode could be apocryphal, many historians believe the Turkish army accidentally attacked itself and killed 10,000 of its own soldiers in the 1788 Battle of Karánsebes during the Austro-Turkish War.

🌐 Since 1945, all British tanks contain a built-in teakettle.

🌐 Warner Bros. Entertainment was founded 3 months before the fall of the Ottoman Empire.

🌐 Ocean liner stewardess Violet Jessop was on board during the 3 largest ship sinkings in history: the *Titanic,* the *Britannic,* and the *Olympic*.

🌐 In 1911, pigtails, or the queue hairstyle, were banned in China because of their connection to the country's feudal history.

🌐

Genghis Khan killed over 40 million people, or 1/6 of Earth's population, during his lifetime.

🌐 Only 4 U.S. presidents can be considered only children: Bill Clinton, Franklin D. Roosevelt, Gerald Ford, and Barack Obama. However, all 4 had half siblings.

In the Enlightenment era, practitioners of the science of cranioscopy believed the shape of the human skull revealed the inner workings of the brain. In the early 19th century, the skulls of Viennese composers Beethoven, Haydn, Mozart, and Schubert were disinterred in order to confirm the existence of a "music bump" that supposedly caused compositional genius.

The first report of strange phenomenon in the Bermuda Triangle was written by Christopher Columbus on October 8, 1492, when the explorer wrote that his ship's compass had stopped working temporarily. Three days later, Columbus and his crew spotted a mysterious light in the distance, causing terror on deck that nearly led to a mutiny.

President Martin Van Buren was born in his family's tavern.

In 1883, a woman tripped on the Brooklyn Bridge. Other people on the bridge thought it was collapsing, then proceeded to run, and 12 people were trampled to death.

The Mongols once celebrated a victory over the Russians by building a stage over the survivors and crushing them to death by having a party atop their bodies.

In Victorian times, bottles were used to collect tears, with a special stopper that allowed the tears to evaporate. When the tears were gone, the mourning period was over.

During the Middle Ages, there was a legal category called "enbrotherment" that allowed 2 men to share living quarters, pool their resources, and effectively live as a married couple.

A "Boston marriage" was a Victorian term for women who lived together without financial support from men. It was used to refer to platonic friendships, but also acted as a euphemism for lesbian couples.

In the Mesozoic period, humans hunted giant armadillo-like creatures and lived inside their shells.

The highest-earning sportsman in history is a Roman chariot racer called Gaius Appuleius Diocles, who earned the modern equivalent of $15 billion.

There is so little archaeological and literary evidence from the period in history between A.D. 614 and 911 that the phantom time hypothesis says those years did not exist at all and that we are currently in the 1700s.

When the pyramids at Giza were being built, there were still isolated populations of mammoths alive in Siberia.

In 1923, German banknotes were so worthless that a fad developed of using them as wallpaper.

Philosopher Tycho Brahe owned a moose, which died after drinking Danish beer at a party and drunkenly falling down the castle stairs.

On the island of Tangier, off the coast of Virginia, inhabitants speak the same language as the 17th-century settlers, which many historians believe is the closest living relative of the Shakespearean accent.

In 1450, a large pack of wolves killed 40 people in Paris.

Damascus, Syria, is widely considered to be the longest continuously inhabited city in the world, with evidence of settlement dating as far back as 9000 B.C.

Bir Tawil is the only area of habitable land not claimed by any nation. An 800-square-mile piece of desert that sits between Sudan and Egypt, it has no settlements and is populated by nomadic tribes who have no allegiance to either country. A Virginian man did try to claim the area so that he could quite literally make his daughter a princess, though his claim is not recognized by the United Nations. The area's unclaimed status is the result of complex border negotiations.

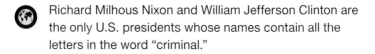 Richard Milhous Nixon and William Jefferson Clinton are the only U.S. presidents whose names contain all the letters in the word "criminal."

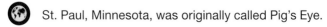 St. Paul, Minnesota, was originally called Pig's Eye.

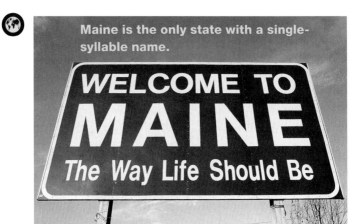

Maine is the only state with a single-syllable name.

WELCOME TO MAINE
The Way Life Should Be

Men's clothes button on the right, but women's clothes button on the left. Historians believe that's because upper-class women used to wear clothing so elaborate that they had to be dressed by other people, hence the mirroring.

The first written instance of "OMG" that we know of was in a letter to Winston Churchill in 1917.

It is legal in 43 states to sue former inmates for the cost of their incarceration.

Idaho's governor, since 2007, is named Butch Otter.

The law that first made public school compulsory in America is called the Old Deluder Satan Law. It was passed by the Massachusetts Bay Colony in 1642.

North Sentinel Island in the Bay of Bengal is home to a tribe that has successfully resisted all attempts at contact from the outside world.

Countess Elizabeth Báthory is believed to be the most prolific female serial killer ever. It's said she murdered about 80 young girls with the help of servants in particularly gruesome ways around the turn of the 17th century. She was spared the death penalty as a Hungarian royal. Her last years were spent in a bricked-up chamber with only small slits for ventilation and the passing of food.

Though One World Trade Center is New York City's tallest building, the city's largest building by floor area is a nondescript office building at 55 Water Street.

The philosopher René Descartes is most well known for the saying "I think, therefore I am," but he also developed the xy-coordinate system.

1. **What was Liberty Island called prior to the Statue of Liberty's arrival?**

   a. Hamilton Island

   b. Wards Island

   c. North Brother Island

   d. Bedloe's Island

2. **Who was the first president to reside in the White House?**

   a. George Washington

   b. Benjamin Franklin

   c. Abraham Lincoln

   d. John Adams

3. **Bras were called what during medieval times?**

   a. Front sacks

   b. Breast bags

   c. Chest guards

   d. Upper cushions

4. **Who was the first African American to win the Nobel Peace Prize?**

   a. Martin Luther King Jr.

   b. Ralph Bunche

   c. Barack Obama

   d. Rosa Parks

5. **At which newspaper did Karl Marx work as a correspondent?**

   a. *New-York Daily Tribune*

   b. *Chicago Sun-Times*

   c. *Times of London*

   d. *Der Spiegel*

6. **Who originated the quote "One death is a tragedy; one million is a statistic"?**

   a. Adolf Hitler

   b. Joseph Stalin

   c. Benito Mussolini

   d. Pol Pot

Art & Literature

Modern art was a CIA plot—the CIA funded American abstract artists in the 1950s and 1960s as part of the culture war against the Soviet Union, as a way of proving the creativity and intellectual freedom of America.

The French playwright Antonin Artaud died clutching a shoe.

A Chinese billionaire bought a $170 million painting by Amedeo Modigliani with his American Express credit card so he could use the points for free airfare.

Charles Dickens had bookbinders print up a number of fake books for his library. Titles included *Drowsy's Recollections of Nothing* (3 volumes), *Hansard's Guide to Refreshing Sleep* (as many volumes as possible), and *Bowwowdom: A Poem*.

- After Charles Dickens's favorite cat died, he turned its paw into a letter opener.

- Congo the chimp was a famous abstract painter in the 1950s, selling paintings to famous artists such as Pablo Picasso and Salvador Dalí. In 2005, one of his paintings sold at auction for almost $26,000.

- Before becoming Pope Pius II, Aeneas Sylvius Piccolomini wrote a popular erotic novel entitled *A Tale of Two Lovers.* It was one of the most widely circulated books of the 15th century.

- As a teenager, Charlotte Brontë and her brother, Branwell, wrote adventure stories in books that measured 1 by 2 inches.

- There's a jetty built on the coast of Croatia that plays music like an organ when waves crash in and out of it.

- One of New York City's smallest museums is found in a freight elevator. Holding mostly everyday objects that are easily overlooked, one of its most famous pieces is the shoe thrown at President George W. Bush at the prime minister's palace in Baghdad.

- The Tate Modern art museum in England purchased 8 million individual sculptures from Ai Weiwei, all handcrafted porcelain sunflower seeds. While each is about the size of a fingernail, altogether they weigh 10 tons. But that's only about 1/10 of the 100 million seeds Ai had made for the original installation.

Vincent van Gogh probably didn't cut his own ear off. Likely, his close friend and fellow painter Paul Gauguin sliced it off with a sword after the two got in an argument, and van Gogh made up the story about cutting it off himself to protect Gauguin. The two never saw each other again after the incident.

The first trombones were called sackbuts, derived from the French words *saquer* and *bouter,* which mean "pull" and "push."

Horror writer Stephen King sleeps with his lights on.

"Preposterous," which these days means "absurd" or "ridiculous," originally referred to mythical animals with heads at both ends, "pre" meaning "front" and "post" meaning "rear."

Tennessee Williams, a severe hypochondriac, died choking on a plastic top from a bottle of nasal spray.

A piano has over 12,000 different parts.

There's music made especially for cats. Apparently, cats develop their musical taste soon after they're born, so cat music includes not only traditional (human-made) instruments, but also feeding noises, bird chirps, and purring sounds.

Playing the violin burns about 170 calories an hour.

Organs were the most complex machines that existed before the Industrial Revolution. The largest organ has 33,114 pipes and weighs 150 tons. Located in Atlantic City, it took 3 years to build. Completed in 1932, it unfortunately is no longer in working condition.

Other names for the soul patch—the small tuft of facial hair directly below the lower lip—include "Dizzy Gillespie beard" and "jazz dab," due to its popularity with jazz and blues musicians.

Most medieval paintings depict babies as small old men. The vast majority of infants were Jesus or other biblical children, and at the time, people thought that Jesus was born perfectly formed and didn't change—he just got bigger. Babies in paintings started to get cuter and look more like babies in the Renaissance, when private citizens started commissioning portraits of their children.

The famous TV painting teacher Bob Ross was an air force sergeant prior to his television career. He had the kind of job that required him to scream at people to make their beds and scrub toilets, and after leaving the military, he decided that he never wanted to scream again.

The first music streaming service available on a phone wasn't an app. Invented in 1897, the Telharmonium transmitted music, played continuously by two keyboard players, over the telephone system. It took up an entire floor of a Manhattan building and its first subscriber was none other than Mark Twain.

When completely stretched out, a trumpet is 6.5 feet long.

The New York Earth Room is an art installation by Walter De Maria that is an entire floor of a downtown New York City loft covered with 280,000 pounds of soil. Built in 1977, the soil has never been replaced, though it is watered once a week. It used to need weeding, but the soil has been barren for some time now.

The designer of the Eiffel Tower built an apartment in the tower itself. Though he didn't live there, he did use it to entertain distinguished guests and scientists.

Karl Marx wrote a novel when he was 19. Never published, it is not considered to be any good by anyone who has read parts of it.

The oldest known piece of jewelry is a necklace of eagle talons made by Neanderthals.

In 1878, James McNeill Whistler sued the art critic John Ruskin for libel after Ruskin claimed Whistler's almost abstract painting of a fireworks explosion was an affront to public taste. Whistler asked the jury to consider the work not as a painting, but as an artistic arrangement. The painting was shown upside down at the trial and Whistler ended up winning, but only a farthing.

The most expensive dollhouse in the world is worth $8.5 million. It is modeled after the castle in Alfred, Lord Tennyson's poem "The Lady of Shalott."

A tiny island in a canal of Mexico City is covered with creepy dolls. The island's caretaker claimed he was haunted by the spirit of a young girl whom he had watched drown and was unable to save. He started placing the dolls on the island to make her spirit happy.

The oldest known wooden sculpture is a 9-foot-tall idol found in a peat bog in Russia that dates back to the agricultural revolution, 11,000 years ago. It was originally 17 feet tall, but nearly half of it was destroyed during the tumultuous 20th century.

The largest piece of pottery in the world is a 5,400-square-foot house made entirely of terra cotta. It has no reinforcing materials and the clay was baked by the sun. Built by Colombian architect Octavio Mendoza over the course of 14 years, the house has 2 stories and looks kind of like a hobbit hole.

The composer Franz Joseph Haydn was expelled from his conservatory for cutting off the ponytail of one of his classmates. Homeless, he sang on the streets for several years.

Another name for brass instruments is *labrosones,* which means "lip-vibrated instruments."

In 1925, members of a Scottish expedition to Antarctica decided to pose for a picture with a penguin while 1 member played the bagpipes. While the penguin looks to be enjoying itself in the photograph, they had to tie it down with a pot filled with snow because it wanted to get away.

C. S. Lewis and J. R. R. Tolkien held contests in which they read bad poetry out loud while attempting not to laugh.

The artist Aelita Andre sold $30,000 worth of paintings in the New York market before she was 5 years old.

Pablo Picasso carried a revolver loaded with blanks, which he would fire at whoever asked him what his work "meant."

Art programs are mandatory in countries whose students score highest in math and science.

Georges Braque was the first living person to have his art displayed in the Louvre.

French sculptor Auguste Rodin died in 1917 from frostbite because he couldn't afford to heat his flat. Meanwhile, his sculptures were housed in Europe's great museums.

French postimpressionist artist Paul Gauguin once worked on the Panama Canal as a construction laborer.

New York School poet Frank O'Hara was killed by a dune buggy on the beach at Fire Island.

American dancer Isadora Duncan was killed when her scarf was caught in the wheel of a car as she sat in the passenger seat, which broke her neck as it drove away. The tragedy prompted Gertrude Stein to quip, "Affectations can be dangerous."

While traveling in Africa in 1954, Ernest Hemingway was injured in 2 successive plane crashes.

Clay pipes containing marijuana residue have been found in William Shakespeare's backyard.

Mexican communist painter Diego Rivera was paid handsomely for his first New York exhibition by capitalist industrialist John Rockefeller.

The epitaph on Los Angeles poet Charles Bukowski's tombstone reads DON'T TRY.

After sneaking into the uncompleted Sistine Chapel to view fellow painter Michelangelo's work, Raphael went back to his own completed painting at the Basilica di Sant'Agostino, scraped it off the wall, and started from scratch.

Bob Ross did *The Joy of Painting* TV series free of charge. All of his earnings came from his art supply store.

Salvador Dalí made a painting for the prisoners at Rikers Island, which hung in the prison dining room for 15 years. In 2003, the painting was stolen by 3 guards and a warden.

The majority of Adolf Hitler's paintings, produced prior to his career as a genocidal dictator, were purchased by Jewish customers.

In 2011, a woman paid $10,000 for a "non-visible" work of art from actor James Franco's Museum of Non-Visible Art.

Serial art vandal Hans-Joachim Bohlmann has caused over $190 million in damages by attempting to destroy famous artworks.

The isotopes strontium-90 and cesium-137 did not exist in nature before the first atomic bomb detonation in 1945. Today, art historians test for these isotopes to determine whether a painting is a forgery. If a work that was supposed to be completed before 1945 contains the post-atomic isotopes, it is a fake.

Schizophrenic artist Mary Barnes developed her talents in a mental hospital, where her therapist, R. D. Laing, encouraged her to paint the walls of her room with her feces. When the smell became too overbearing, she was given paint.

Vincent van Gogh painted *The Starry Night* in an insane asylum in Saint-Rémy-de-Provence, France. The painting is a depiction of the artist's view from his room. The asylum now has a wing named for van Gogh.

Artist Peter von Tiesenhausen made the top 6 inches of his 800-acre farm a copyrighted work of art to keep oil pipelines out. It costs oil companies $500 an hour to talk to him.

The first published book is believed to be the *Epic of Gilgamesh,* written in 3000 B.C.

The first published novel is believed to be *The Tale of Genji,* written in the early 11th century.

Ernest Wright's 1939 novel *Gadsby* does not contain the letter "e."

In the original Italian children's story, Pinocchio squashes Jiminy Cricket to death.

 Leonardo da Vinci's late 15th-century mural painting *The Last Supper* originally showed Jesus's feet, but they were removed from the image during a renovation in 1652 to accommodate a doorway on the wall the painting was hanging on.

 Dutch painter Piet Mondrian frequently worked on his paintings for so long that his hands blistered.

 Johannes Vermeer used a camera obscura while sketching his paintings.

 During World War I, playwright Bertolt Brecht worked as a medical orderly in a clinic for treating soldiers infected with venereal disease.

 Upon obtaining a university degree in 1839, the poet Charles Baudelaire wrote to his brother that he did not have a vocation for anything.

Sculptor Louise Bourgeois's father had a long affair with her nanny.

Filmmaker Kenneth Anger attended the Santa Monica cotillion in his youth, which encouraged famous and non-famous children to interact. It was there that Anger had his first dance, with Shirley Temple. Anger's early work would later be the subject of an obscenity trial, and he would tattoo the name "Lucifer" across his stomach.

Werner Herzog once promised to eat his own shoe if Errol Morris could actually complete the documentary on pet cemeteries he'd been working on for years. *Gates of Heaven* premiered in 1978 to international acclaim, and Herzog kept good on his word, resulting in the creation of a film called *Werner Herzog Eats His Shoe*.

While living in Baltimore, Maryland, F. Scott Fitzgerald was hospitalized 9 times at Johns Hopkins Hospital for injuries sustained because of or during drinking.

Early in his career, Irish novelist James Joyce was found drunk and beaten on St. Stephen's Green by a friend of his father's. This man, Alfred Hunter, would be the inspiration for Joyce's most famous character, Leopold Bloom in *Ulysses*.

Plato believed that because poetry was irrational, poets should be outlawed in the ideal city-state.

Visionary poet and painter William Blake was walking in London when he got caught up in an angry mob attempting to storm Newgate Prison, an event which became known as the Gordon Riots and which sparked the creation of the first police force in Britain.

Poet Philip Larkin never left England. When asked why in an interview, Larkin responded, "I never saw the point."

The two most famous poets of the romantic era, William Wordsworth and Samuel Taylor Coleridge, were close friends until Wordsworth ended the relationship due to Coleridge's growing opium habit—a habit that would inspire his most famous work, *Kubla Khan*.

While living in the Netherlands, Claude Monet purchased a small boat and built a studio on top of it so he could explore the local lakes and lilies.

Rembrandt's grave and remains were destroyed 20 years after his death, as was customary for the burial places of the poor.

The patronage system in the Philippines is locally known as the star fruit system.

Fyodor Dostoyevsky's reclusiveness and religiosity as a youth earned him the nickname "Monk Photius."

By 1949, novelist and playwright Jean Genet had been found guilty of indecency, vagrancy, and theft so many times throughout his life that he was facing a life sentence in prison. Fortunately for world literature, prominent cultural figures, including Jean Cocteau, Pablo Picasso, and Jean-Paul Sartre, signed a petition that led to Genet's release, and he never did time in prison again for the rest of his life.

Legendary Swedish filmmaker Ingmar Bergman admitted in an interview that his own films depressed him and he could no longer watch them.

Psychedelic Chilean filmmaker Alejandro Jodorowsky was a circus clown before he began working in the visual arts.

Paint made from artificial resins was invented because of linseed oil shortages during World War II.

In 1939, William S. Burroughs deliberately severed the tip of his pinkie finger to impress a man he was infatuated with.

📖 There are over 1,000 adaptations of Shakespeare's works.

📖 *The Red Vineyards near Arles* is the only painting by Vincent van Gogh to be sold while the artist was still alive.

📖 The *Mona Lisa* by Leonardo da Vinci is most likely a portrait of Lisa del Giocondo (née Gherardini), wife of a rich Florentine silk merchant. It was commissioned for the family's new home after the birth of their second child, sometime around 1505.

📖 Walt Whitman's brothers were named George Washington Whitman, Thomas Jefferson Whitman, and Andrew Jackson Whitman.

📖 Walt Whitman started 2 newspapers in his lifetime: the *Brooklyn Weekly Freeman* and the *Long Islander.* The *Long Islander* is still in print today.

📖 Mary Shelley's mother, Mary Wollstonecraft, was one of the most prominent anarchist and feminist thinkers of her day. She wrote one of the first feminist tracts, *A Vindication of the Rights of Woman.* She died just days after giving birth to Shelley.

📖 Don José Ruiz Blasco, Pablo Picasso's father, was an artist who specialized in depicting animals and other natural scenes.

The idea for Mary Shelley's *Frankenstein* came about after discussing the occult with fellow writers Percy Bysshe Shelley and Lord Byron. They decided to have a horror-story writing competition, and Mary Shelley ultimately spun her idea out into a full-length novel.

American Gothic by Grant Wood is modeled after the artist's sister and dentist. Grant wanted to use his mother but determined that standing for the painting would be too exhausting, so he had his sister wear her apron and brooch.

Pablo Picasso's last words were supposedly "Drink to me, drink to my health, you know I can't drink anymore." He died April 8, 1973, in Mougins, France.

Oscar Wilde's alleged last words were "My wallpaper and I are fighting a duel to the death. One or the other of us has to go."

Always his own harshest critic, Leonardo da Vinci's last words were "I have offended God and mankind because my work did not reach the quality it should have."

Wolfgang Amadeus Mozart's last words were "The taste of death is upon my lips. . . . I feel something that is not of this earth."

Vlad Dracula ~
the Impaler

The Dracula legend was inspired by 15th-century Romanian ruler Vlad the Impaler, who is said to have soaked his bread in the blood of slaughtered enemy soldiers.

Dr. Seuss (Theodor Seuss Geisel) had a massive collection of funny hats, kept in a room hidden behind a bookshelf in his home. He would put them on to overcome creative blocks, or have dinner party guests wear them and request each guest "play" the "character" of their respective headpiece.

John James Audubon made it his life's mission to paint every bird in North America. Between 1827 and 1838, he painted 435 life-size watercolors of North American birds.

Artist Damien Hirst directed the music video for the song "Country House" by the Britpop band Blur.

Jane Austen was the youngest girl of 8 children. She began writing early as a way to entertain her siblings and was incredibly prolific from a young age. She completed full drafts of *Northanger Abbey, Sense and Sensibility,* and *Pride and Prejudice* by age 23.

Claude Monet's father disapproved of his painting and encouraged him to become a grocer.

Pablo Picasso's full name is Pablo Diego José Francisco de Paula Juan Nepomuceno María de los Remedios Cipriano de la Santísima Trinidad Ruiz y Picasso.

Auguste Rodin's statue *The Age of Bronze* was such a realistic depiction of the human form that the artist was thought to have entombed an actual person in bronze. Ultimately the controversy surrounding the statue was a boon to Rodin as people flocked to see his fabled work. He was able to sell casts of the statue to museums in London, New York City, Tokyo, Berlin, and elsewhere.

Flemish painter Peter Paul Rubens was knighted by both Philip IV of Spain and Charles I of England.

Toni Morrison was the first African American woman to receive a Nobel Prize. It was in recognition of her contributions to literature and poetry.

British artist Willard Wigan creates microsculptures so small, you need a microscope to see them. His work often sits in the eye of a sewing needle or on the head of a pin. He got his start at 5 years old making houses for ants because he thought they needed a place to live.

Åke "Dacke" Axelsson hung 4 paintings done by a chimpanzee during a 1964 avant-garde art exhibition in Sweden, under the pseudonym Pierre Brassau. Critics praised the works, even after they discovered the artist was a chimpanzee.

Polish artist Rafał Bujnowski submitted a photo of a hyperrealistic self-portrait he had painted to the U.S. embassy as part of his visa application. The painting was so photorealistic that no one in the visa office even noticed! He later received a U.S. passport with his self-portrait as the picture.

British priest Father Jamie MacLeod bought a dusty painting at a Cheshire antiques shop for £400. He took it on the British TV show *Antiques Roadshow* and discovered it was a rare Sir Anthony van Dyck portrait worth £400,000. He used the money from the sale of the painting to buy new bells for his church.

In 1955, monks at a Buddhist temple were moving an ancient Buddha statue made of plaster. They dropped it and the plaster shattered, revealing a solid gold statue underneath.

Artist David Choe accepted stock as payment when he painted a mural at Facebook's Silicon Valley office. His stock was valued at $200 million during Facebook's 2012 IPO.

Auguste Rodin's famous sculpture *The Thinker* was originally only about 70 centimeters high and called *The Poet*. It was meant to be a small part of a larger sculptural representation of Dante's *Inferno* that Rodin was working on in the 1880s. *The Poet* was supposed to represent Dante himself. The larger, more popular version we know today wasn't created until 1904.

 Pietà is the only known work of art to actually be signed by Michelangelo. It is a marble sculpture of the body of Jesus, after crucifixion, in Mary's lap. It is housed in St. Peter's Basilica in Vatican City.

Pablo Picasso was hauled in for questioning when the *Mona Lisa* was stolen in 1911. He was deemed innocent and released, though there is still some speculation as to his involvement. Prior to the theft, the painting was barely known outside of the intelligentsia. Its current fame owes much to media sensationalism.

The earliest known art—African cave paintings and stone carvings—dates back to roughly 32,000 B.C.

Georgia O'Keeffe was somewhat annoyed at the publicity surrounding her flower paintings. She told art critic Emily Genauer: "I hate flowers—I paint them because they're cheaper than models and they don't move."

Artist Ivan Albright was so meticulous, he often worked with a single-haired brush and would spend whole days working on 1 square inch of canvas.

In 1961, the Henri Matisse painting *Le Bateau* ("The Boat") hung upside down for 47 days in the New York Museum of Modern Art before a patron, stockbroker Genevieve Habert, noticed the error and alerted the staff.

April 15 is World Art Day and Leonardo da Vinci's birthday.

E. B. White, famous children's book author of *Stuart Little* and *Charlotte's Web,* is also the coauthor of William Strunk Jr.'s *The Elements of Style,* one of the most widely used American English writing style guides.

Aldous Huxley, author of *Brave New World* and *Island,* wrote the original screenplay for Disney's adaptation of *Alice's Adventures in Wonderland.* Walt Disney rejected the script, finding it too literary and complex.

Joseph Stalin loved westerns and his favorite actor was John Wayne. What he wasn't a fan of in film was romance, even banning kissing in Russian-made films for a period.

The phenomenon of copycat suicides or suicide clusters is called the Werther effect, after Goethe's novel *The Sorrows of Young Werther.*

In 2013, Russian conceptual artist Petr Pavlensky nailed his scrotum to the ground in Moscow's Red Square, calling the act a metaphor for the apathy of contemporary Russian society.

George Orwell, author of *1984* and *Animal Farm,* once published a very detailed essay on how to make the perfect cup of tea. It was titled "A Nice Cup of Tea."

📖 "Boatswain"—the word for the senior crewman responsible for the components of a ship's hull—is pronounced *bosun*.

📖 "Abstemiously" and "facetiously" are the only English words with all the vowels in alphabetical order.

📖 A British woman held hostage on a Maoist commune for 30 years by her father was inspired to escape after reading the Harry Potter and *Lord of the Rings* novels.

📖 The word "huh" exists in every language on the planet.

📖 Mozart's full name was Johannes Chrysostomus Wolfgangus Theophilus Mozart.

📖 The only word that rhymes with "purple" is "hirple," which means "to limp awkwardly." Nothing rhymes with "woman."

📖 The word "calf," referring to a baby cow, is of Germanic origin, whereas "calf," referring to the lower leg, comes from Old Norse.

📖 Most chiming clocks play the same tune, called "Westminster Quarters." It's named after the Palace of Westminster, where Big Ben hangs.

Moby-Dick was inspired by a true story. The real-life Ahab was George Pollard, captain of the whaling ship *Essex*. The *Essex*'s crew was responsible for the extinction of 2 tortoise species. When a killer whale rammed the ship, the crew ended up stranded in 20-foot longboats, ultimately resorting to cannibalism. The survivors were found delirious, sucking on the marrow of their dead compatriots' bones.

Eigengrau is the German word for the color you see when there's no light. It means, literally, "own gray."

Leonardo da Vinci was born a bastard to unwed parents.

Salvador Dalí believed he was his dead brother's reincarnation. His brother had also been named Salvador and died before the painter was born.

There is either a portrait or a silhouette of Salvador Dalí in all of his paintings.

The most recognized portrait of Priapus, the Greek god of fertility, features a prominent penis with clear signs of shut phimosis, which would have rendered him infertile.

Michelangelo di Lodovico Buonarroti Simoni's father, Lodovico, beat his son when he first began to draw, considering it inappropriate for a man. He was reportedly mortified later when his son became an artist by vocation.

- Ultramarine was at least as valuable as gold during the Renaissance. An ounce of the pigment cost about $5,000 in contemporary value and was typically used only for painting the most sacred artistic subjects, such as the Virgin Mary's robes.

- In 2015, Paul Gauguin's painting *Nafea Faa Ipoipo* ("When Will You Marry?") sold privately for $300 million, making it the most expensive painting in history.

- Rembrandt's *The Conspiracy of Claudius Civilis,* completed in 1662, was the artist's largest work at 18 feet across. Due to financial straits, Rembrandt was forced to cut the painting down to a quarter of its size and sell it. Today, it hangs in Sweden's National Museum of Fine Arts.

- The first color photograph was taken in 1861 by Thomas Sutton, using James Clerk Maxwell's three-color method. The subject of the photo is a tartan ribbon.

- The contemporary art form of leaf carving involves the meticulous removal of a leaf's surface to produce a beautiful design on the leaf, almost like a silhouette. It takes leaf artists 3–4 months on average to complete a single work.

- Masonite, a popular support material for oil paintings, was patented in 1924. Masonite board is made by shooting wood chips into tubelike fibers filled with steam, then heating and molding the by-product.

The Mozart effect refers to the phenomenon of increasing IQ by listening to Wolfgang Amadeus Mozart's compositions. The idea was introduced and coined by Don Campbell in 1997. It led to the sale of Baby Mozart albums, and in 1998, the governor of Georgia allotted $105,000 in the state budget to ensure all children had access to Mozart's music. Today, most psychologists consider the Mozart effect to be a myth.

Leonardo da Vinci painted in ultrathin microglazes, 50 times thinner than a human hair, and layered his works extensively. He painted about 30 layers on the *Mona Lisa*.

Italian Renaissance artists used red lac pigments to represent human flesh. Red lac is made from parasitic insects that live on the kermes oak.

Since 1974, the *Mona Lisa* has been stored in a bulletproof, climate-controlled glass case with a bed of silica to keep the air inside at 55% humidity.

Michelangelo painted 300 characters on the ceiling of the Sistine Chapel, and no two are even remotely alike.

Having renounced his entire body of work all before his 21st birthday, the French symbolist poet Arthur Rimbaud embarked on a second career as an adventurer. In 1884, he became a coffee and weapons merchant in Harar, Ethiopia, where his best friend was Governor Ras Makonnen, father of future Ethiopian emperor Haile Selassie.

Unbeknownst to most people, there are actually 3 versions of Leonardo da Vinci's *Mona Lisa*. In addition to the most famous version hanging in the Louvre, a second verified version is in Madrid's Prado museum. The third, however, known as the *Isleworth Mona Lisa,* is believed to be a decade older than the first two, and despite extensive analysis by art historians, it remains unclear whether the *Isleworth* portrait was painted by da Vinci himself or by one of his students.

In May 2010, performance artist Marina Abramović spent 736 hours in the Museum of Modern Art staring at museumgoers, face-to-face, as part of her performance retrospective entitled *The Artist Is Present.*

At the 1900 Paris Olympics, winners received valuable works of art instead of gold medals.

In 2012, the Smithsonian officially recognized video games as an art form and had an exhibit to "comprehensively examine the evolution of video games as an artistic medium."

There is a record entitled *The Best of Marcel Marceao* [*sic*] that is nothing more than 38 minutes of complete silence followed by thunderous applause. Marcel Marceau was a mime.

The artist Man Ray created an object in 1923 called *Object to Be Destroyed*. It survived until 1957, when a group of nihilists stole it from an exhibition and shot it.

In 1988, artist Teemu Mäki made a video in which he killed a cat and masturbated on its corpse. He was the professor of fine arts at Aalto University in Finland from 2008 to 2013.

In 1961, Italian artist Piero Manzoni defecated into 90 tin cans, each containing 30 grams of feces. A single can is worth €100,000 today. He has also sold his breath.

The artist Max Neuhaus's installation *Times Square* is hidden under a subway grate in Manhattan and emits a loud, ominous hum audible to those walking over it.

Before he was famous, Pablo Picasso sometimes burned his own paintings to keep warm.

Benjamin Franklin once wrote a lengthy essay addressed to the Royal Academy of Arts titled "Fart Proudly."

The 1959 cinematic epic *Ben-Hur* was a remake of a 1907 film of the same name, which was adapted from an 1880 novel called *Ben-Hur*.

English is the only Indo-European language that isn't gendered.

Kenneth Goldsmith "wrote" a poem called "The Day" in which he typed out an entire day's *New York Times* content. It took him a year to complete and is 836 pages long.

The Icelandic language has changed so little in the past thousand years that a modern-day Icelander would be able to understand Old Norse. Meanwhile, try reading *Beowulf* in Old English: "Hwæt! We Gar-Dena in geardagum þeod-cyninga, þrym gefrunon, hu ða æþelingas ellen fremedon!" Thought so . . .

The first artists were mostly women, according to archaeologists. Analysis of handprint stencils on prehistoric cave paintings—which most likely served as signatures—has shown that 3/4 of them were left by women. Although an alternate hypothesis is that they were left by adolescent boys.

Many words that relate to art, culture, and government entered the English language when the Normans—who were French—ruled England. At the time, many people considered Latinate words to be pretentious. And that remains the reason English speakers tend to think that long words are fancier. Latin-based words are generally longer than Germanic words.

English is the only known language on the planet whose present tense requires a special ending in the third person singular. For instance: "I write, you write, she writes."

John Cage's most famous composition is "4'33"," in which musicians stand onstage and don't make a sound for 4 minutes and 33 seconds. The composition is actually made up of the noises of the audience.

A Leonardo da Vinci painting worth $150 million may actually be a forgery inspired by a grocery store clerk.

Richard Prince sold "paintings" that were screenshots of him commenting on other people's Instagram accounts—usually those of attractive young women. Collectors snapped up the works at $100,000 apiece.

The poet Gaius Helvius Cinna was torn apart by a mob after the assassination of Julius Caesar, a case of mistaken identity. The mob thought he was Cornelius Cinna, one of Brutus's coconspirators.

The avant-garde ballet *The Rite of Spring* nearly caused a riot when it was first performed in Paris in 1913. The cacophonous music and stomping and lurching of the dancers caused violent arguments to break out among the audience, and people started to throw things at the orchestra; 40 people had to be ejected before the ballet could resume.

There is no one specific word for "blue" in the Russian language.

1. **What author, with over 1,000 adaptations and counting, is history's most adapted author?**

    a. Tom Clancy

    b. Stephen King

    c. William Shakespeare

    d. Victor Hugo

2. **Which artist was encouraged by his father to stop painting and become a grocer?**

    a. Claude Monet

    b. Pablo Picasso

    c. Leonardo da Vinci

    d. Andy Warhol

3. **Artist David Choe accepted stock as payment when he painted a mural at this company's Silicon Valley office. His stock was valued at $200 million during the IPO.**

    a. Facebook

    b. Twitter

    c. Snapchat

    d. Instagram

4. **Who was the first living person to have his art on display at the Louvre?**

    a. Pablo Picasso

    b. Georges Braque

    c. Henri Matisse

    d. Jackson Pollock

Health & Medicine

Napping before you get tired is actually more effective than napping when you're already sleepy.

The brain uses 20% of your body's oxygen even though it makes up only about 2% of your overall body weight.

Men with deep voices tend to have lower sperm counts than men with higher-pitched voices.

Bananas are more effective in replenishing electrolytes than Gatorade. They also have serotonin and dopamine—chemicals that help you feel happy.

1 in 10 Americans has a lifelong drug addiction of some kind.

Women get drunk quicker than men when drinking the same amount of alcohol because men have more water in their bloodstreams, which dilutes the alcohol.

San Franciscans have slept with more people than the rest of the country, averaging 30 sexual partners per person. But Los Angeles residents have more sex than everyone else, on average 135 times a year.

98% of the atoms in a person's body are replaced each year.

Coffee reduces the risk of most cancers.

Research has suggested that watching pornography could make men better weightlifters.

Castrated men tend to live longer.

Almost all infants in Finland spend their first months sleeping in a cardboard box. Finnish mothers are given the option of a small stipend or a box filled with baby essentials such as clothes and bedding, with the box doubling as a bassinet.

Snot from a sneeze can travel up to 200 mph.

It's quite rare, but broken heart syndrome can literally kill you.

Research has shown that people prefer the smell of their own farts to those of other people.

The G-spot is named after the gynecologist who first discovered it, Ernst Gräfenberg, but it was almost called the Whipple Tickle after the sex researcher Beverly Whipple.

Most fat leaves your body through your breath.

An Israeli woman won a $1,000 lawsuit against a weatherman after he predicted sunny weather. She claimed the forecast caused her to leave the house lightly dressed, but it rained and she caught the flu.

- The risk of medical complications increases in July, when new residents enter hospitals.

- Dr. Jerri Nielsen successfully treated herself for breast cancer while stranded in Antarctica. However, the cancer returned 10 years later and she died at the age of 57.

- Due to a gene mutation that occurred in the 1700s, 38 people in a small town in northern Italy don't suffer from cholesterol artery-clogging. They are immune to heart disease and stroke.

- Wood chopping boosts testosterone production by almost 40%.

- Islamabad, Pakistan, is the hay fever capital of the world.

- Low wages are strongly linked to the medical condition hypertension.

- An 8-week meditation course causes the amygdala, associated with fear and other emotions, to shrink while the prefrontal cortex, associated with awareness, concentration, and decision-making, will thicken.

- Autoimmune encephalitis is a rare neurological disorder that affects mostly women and closely mimics the symptoms associated with demonic possession.

Poor exercise habits kill about as many people as smoking tobacco.

Researchers believe rosacea is caused by mites that live and mate on your facial skin.

The Mariko Aoki phenomenon is the urge to defecate after smelling books.

Cuddling releases oxytocin, which can significantly enhance the healing of wounds.

The average person can survive for 5.5 hours in a buried coffin before running out of oxygen. The smaller the person, the longer they're likely to survive because less body mass would take up the space of air in the coffin.

Multiple sclerosis occurs far less in tropical and semitropical climates than cold climates. While nobody knows why this discrepancy exists, one theory asserts that it could be due to the higher percentage of dairy products consumed in colder regions than warm ones.

While artificial organs and limbs are widely available, there are not, as of yet, any artificial oxygen-carrying blood substitutes. However, there are some non-blood-based substitutes that expand the volume of blood, but they do not allow for the passage of additional oxygen.

- 30% of pregnant women experience cravings for nonfood items, an eating disorder known as pica. The most common substances craved during pregnancy are dirt, clay, and laundry starch.

- Phobophobia is the fear of having a phobia. Symptoms include dizziness, excessive sweating, increased heart rate, and faintness.

- 21% of people in France have experienced a major depressive episode, making it the most depressed nation in the world. The United States is the 2nd most depressed country.

- There have been 40 recorded cases around the world of people suffering from the Truman Show delusion—a disorder named after the 1998 film starring Jim Carrey—in which victims believe their lives are staged plays or reality shows.

- A drop of human blood travels 60,000 miles through the body every day.

- The lungs inhale more than 500,000 gallons of air each day. If you were to "unroll" the lungs, their surface area would cover half a tennis court.

- **A person can expect to breathe in about 45 pounds of dust in their lifetime.**

- Kissing for 1 minute burns 26 calories on average.

Every time you lick a USPS stamp, you ingest about 10% of a calorie. British stamps, however, contain about 5.9 calories. Israeli stamps are kosher.

The average human eyelash lasts 150 days.

The average human sloughs off 8.8 pounds of skin cells per year.

Perspiration is actually odorless. The smell is caused by bacteria that live on skin.

Women blink 19 times per minute on average, while men blink 11 times per minute.

Gay men blink more than straight men and lesbians blink less than straight women.

It takes just 7 pounds of pressure to rip off a human ear.

- Out of 1,000 surgeries, 1–2 people wake up during surgery while under anesthesia. Terrifyingly, children are 8–10 times more likely to experience anesthesia awareness.

- Standing for 3 hours each day at work instead of sitting uses the equivalent amount of energy as running 10 marathons a year.

- Intelligent people generally live longer, and research suggests that there's a genetic factor, rather than environmental. However, scientists have not been able to figure out exactly what the reason is.

- The death rate for babies born on the weekend is 7% higher than those born during the week.

- The town of Cândido Godói in Brazil has a twin birth rate 18 times higher than normal. It was claimed that the infamous Nazi doctor Josef Mengele caused a mutation in the population while experimenting on women after he fled there following the fall of Nazi Germany. However, doctors have concluded it's most likely a result of inbreeding in the small German enclave.

- Women burn more calories during sex if they orgasm.

- In the 1960s, doctors tried to power artificial hearts with plutonium. It didn't catch on, since plutonium is radioactive, though there are still a number of patients alive with plutonium-powered hearts.

A hot drink on a hot day can cool you down. Drinking a hot beverage makes you sweat, and when the sweat evaporates, your body temperature falls.

The corners of your mouth crack because of an infection caused by fungus or yeast.

An industrial designer developed a 3-D-printed cast that uses ultrasound to heal bones 40% faster. It's made in a lattice pattern that provides ventilation, so it doesn't itch or smell like traditional casts.

Only 1 person has ever been cured of HIV. Timothy Ray Brown had been HIV positive for 11 years when he was diagnosed with a rare leukemia. Brown received 2 bone marrow transplants and the virus disappeared completely. Scientists still don't know why exactly it happened.

Chaga, a fungus found on birch trees only in winter, has been shown to have curative properties, particularly against stomach ailments, and can possibly prevent some types of cancer.

When New York City's Grand Central Station was restored in 1998, 1 brick in the ceiling was left untouched. The grime covering it is 70% tar and nicotine from cigarette smoke. You can find it in the zodiac in the main concourse, right next to the constellation Cancer.

- Researchers have been able to pin down the origin of HIV to Kinshasa, the capital of the Democratic Republic of the Congo, in the 1920s. With an extensive rail network, Kinshasa was one of the most connected cities in Africa at the time, making it easy for the virus to rapidly spread all over the world.

- Red hair was far more common in Neanderthals than in humans.

- Long-distance runner Dean Karnazes ran 50 marathons in 50 states in 50 consecutive days. Karnazes's body doesn't break down like most people's do, and theoretically he could run at a 10-minute-mile pace indefinitely as long as he got enough food and water.

- A Cleveland brain surgeon successfully transplanted the head of a monkey onto another monkey.

- Misophonia is a neurological disorder defined by getting annoyed when other people eat around you.

- Hair outside the skin is already dead and doesn't really need to be washed to prevent it from smelling bad.

- Witzelsucht is a neurological disorder in which people make constant bad puns in inappropriate situations.

Hungarian physician Ignaz Semmelweis discovered that by washing his hands between delivering babies, he cut infant mortality rates in his practice by 35%.

In 1976, a swine flu outbreak in the United States killed 1 person, but the swine flu vaccine killed 25 people.

Kissing is more hygienic than shaking hands.

You can develop leg paralysis permanently by sitting on your wallet too often.

Expecting mothers suffering from high stress tend to give birth to shorter children.

Having little social interaction is as harmful as smoking 15 cigarettes per day.

Honey is a better cough suppressant than over-the-counter cough suppressants.

In 1936, a doctor intentionally overdosed on cocaine, and as he was dying, he wrote a detailed biological description of the experience on the wall of his office for other scientists to read.

Putting raw sugar on a cut will stop the flow of blood.

In the past decade, more U.S. citizens have died from prescription painkiller overdoses than heroin and cocaine overdoses combined.

The Bristol stool scale was developed to classify forms of human excrement into 7 categories.

Using a basic food-reward system, normal household dogs were trained to detect both lung and breast cancer by smelling patients' breath, with a 90–100% success rate when checked by biopsy confirmation.

Bootleggers during Prohibition added an agent to their "patent medicine" that paralyzed its users partially to fully; there were approximately 50,000 victims. Those able to walk had a distinctive gait known as jake leg.

As recently as 1985, infants and children up to 18 months old were routinely operated on without anesthetic because it was believed that they could not feel pain.

In 2001, Philip Morris's Czech division published a report saying cigarettes saved the government money because smokers died before retiring.

After leaving office, President Jimmy Carter started an effort to eradicate the guinea worm, which infected millions of people with worms that slowly and painfully burrow out of the body over 3 months. By 2001, the infection rate was reduced by 98%, with 80% of the remaining cases in the Sudan war zone.

President Franklin D. Roosevelt founded an organization to find a cure for polio, and he believed that if every American gave only a dime, polio would be eradicated. Because of this motto, after his death in 1945, FDR's face was put on the dime, and his organization was renamed the March of Dimes.

- Scientists re-created a 1,000-year-old onion-and-garlic eye remedy from an Anglo-Saxon manuscript and found that it killed 90% of antibiotic-resistant staph bacteria (MRSA).

- Kissing for 30 minutes can reduce the effect of allergic reactions among atopic and allergic patients.

- Fraudulent doctor John R. Brinkley replaced the testicles of 16,000 impotent men with goat testicles in the early 20th century.

- If you pour cold water in a person's ear, their eyes will move in the direction of the opposite ear. If you pour warm water in a person's ear, their eyes will move in the direction of that ear.

- Pornography-induced erectile dysfunction is a growing problem for young adult males.

- In 1976, Los Angeles doctors went on strike and the city's mortality rate dropped 18%.

- Oaxacan woman Inés Ramírez Pérez performed a caesarean section on herself with just a few shots of tequila as an anesthetic.

- Diane Van Deren, one of the world's top ultra-runners, lacks the ability to form new memories, which she says is an advantage since she never knows how far she has left to go when racing.

In 1988, an African woman who was born without a vagina gave birth. Apparently, 278 days earlier, she'd been treated for a stab wound when her former lover caught her with a new boyfriend. Somehow, semen got on the knife that entered her abdomen and impregnated her.

In Germany there are fake bus stops outside of nursing homes to prevent senior citizens with dementia from wandering away.

In 1984, Pope John Paul II kissed several lepers on the forehead to prove that leprosy is not as contagious as commonly thought.

A woman in Memphis, Tennessee, is the only person in the world with a condition that causes her to grow fingernails out of every hair follicle on her body.

A woman with polio was kept alive in an iron lung for 58 years, but died when a power outage shut down the machine.

Continuum magazine was founded with the singular aim of denying the existence of AIDS. It folded after all the editors died of AIDS-related complications.

In 2008, the government of Japan prescribed a waist-size limit for all males in the country: 33.5 inches.

- In Goa, health officials are concerned about the prevalence of so-called pig toilets—farms where pigs are fed human excrement.

- A cat's purr can increase its bone density and ability to heal from injuries.

- Super Glue can effectively heal wounds and actually saved the lives of many American soldiers during the Vietnam War.

- In 1883, German created the world's first universal health-care system.

- Daniel Kish is a blind man who uses echolocation to ride his bicycle in traffic.

- In 1989, a child was born to a mother who'd been kept on life support for 107 days.

- In 1964, San Diego high school student Randy Gardner stayed awake for 11 days straight. His world record remains uncontested.

- Every citizen of Portugal is an organ donor by default.

- A 40-year-old man developed pedophilic urges caused by a brain tumor. When the tumor was removed, so were the urges.

The risk of serious heart attack increases by 4% in the first 3 weeks after daylight saving time.

King Zog I of Albania smoked 225 cigarettes a day and survived 55 assassination attempts.

It takes 20% more anesthesia to knock out a redhead than a person of another hair color.

The average American family spends more money annually on soft drinks than water.

Shaolin monks conditioned their testicles to withstand heavy beatings by tying heavy objects to their scrotums and lugging them around.

Semen contains a natural antidepressant.

Cold weather can burn more calories and reduce inflammation.

- Only 5–10% of people who undergo CPR survive.

- Low levels of cholesterol have been associated with violence, aggression, and suicide in several studies.

- Dopaminergic drugs, which change the way the body produces and reacts to dopamine, are used to treat a wide variety of conditions, including restless legs syndrome and Alzheimer's, and can cause compulsive gambling as a side effect. Although this side effect is not widely understood, it is believed to be caused by the brain's reward and pleasure system, which releases the chemical compound dopamine.

- Increases in safe sexual activity for postmenopausal women have been linked to healthier vaginas.

- The largest internal organ is actually the small intestine. If it was removed from your body and straightened out, it would be over 20 feet long.

- **Your fingernails grow faster on your longer fingers. Middle fingers, which are the longest, have the fastest fingernail growth.**

The average erect penis is 5.56 inches long according to a 2013 survey by the *Journal of Sexual Medicine*. The survey included about 1,600 men.

The clitoris has over 8,000 nerve endings, more than double that of the penis.

Heart attacks are statistically more likely to occur on a Monday. Although it is not entirely clear why, many people blame the stress of starting the workweek.

You cannot laugh by tickling yourself. The laughter reaction to tickling has to be caused by external stimuli.

Nerve impulses moving through the brain travel at a speed of about 170 mph.

Fear causes the human body to produce more earwax. The cause of this remains unknown.

Humans are the only animals known to produce tears due to emotional responses.

Women burn on average about 50 calories less than men a day, even when accounting for different levels of activity.

Hydrochloric acid, the kind found in your stomach, is strong enough to dissolve stainless steel.

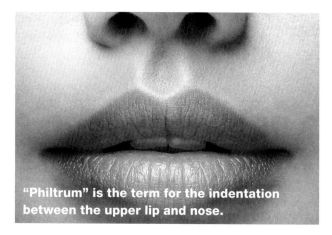

"Philtrum" is the term for the indentation between the upper lip and nose.

The average person loses 60–100 hair follicles a day. The exact number is a product of age, health, diet, and gender.

Fingernails grow 4 times faster than toenails.

Christmas Day, the day after, and New Year's Day are the most common days for heart attacks.

Women's hearts tend to beat faster than men's hearts. This is because women tend to be smaller than men and have less body mass to supply with blood. Men also tend to have larger arteries. The differences between women's and men's hearts are pronounced enough that they need to be accounted for when treating cardiovascular disease.

People tend to think of their liver only after a night of heavy drinking, but the liver has over 500 vital functions. Not only does it detoxify the body, but it also decomposes damaged red blood cells and helps synthesize certain proteins—just to name a few functions.

The left lung is slightly smaller than the right lung to make room for the heart.

When a sick person coughs, he can spread germs at up to 60 mph.

The average person farts 14 times a day.

On any given day, roughly 4% of the world's population will have sexual intercourse. That comes out to about 120 million people.

The largest single cell in the human body is the ovum while the smallest is the sperm cell.

Roughly 1 in 2,000 babies is born with at least 1 tooth.

Newborns tend to have light-colored eyes because it takes a little while for the body to start producing melanin and other eye pigments. It is not uncommon for a baby to be born with green or blue eyes and then to develop brown eyes a few weeks after birth.

You acquire your unique, lifelong fingerprint when you are just 3 months old in the womb.

Only about 33% of the world's population has 20/20, or "perfect," vision.

Hair and nails do not continue to grow after we die. That myth came to be because in death the body dehydrates and the skin tightens, revealing more of the hair's root or the base of the nail, giving the illusion of growth.

We lose some of our tasting abilities as we grow older. By age 60, most people have roughly half the taste buds they had as children.

Babies are born with roughly 300 bones. But because many bones fuse together over time, the average adult has 206 bones. About 1/4 of those bones are in your feet.

We are roughly 1 centimeter taller in the morning. Walking, sitting, and other activities that we do during the day compress our cartilage. At night it decompresses and we wake up slightly taller than when we went to bed.

Simply taking 1 step uses over 200 muscles in the body.

By some measurements each inch of your skin is home to over 32 million bacteria.

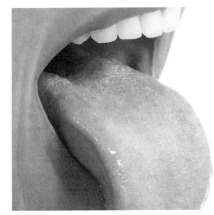

Along with a unique fingerprint, every adult also has a unique tongue print.

There is enough iron in your body to make a small coin.

The world's most common blood type is O, which can be given to people with both type A and type B blood. The rarest blood type in the world is A-H, or Bombay blood, which is thought to occur in only a few hundred people.

Goose bumps are thought to be a remnant from an earlier evolutionary time when humans had more hair on their bodies.

Your skin will completely replace itself about 900 times during your lifetime.

In the early days of dentistry, false teeth often came from the corpses of deceased soldiers.

- Until recently, scientists didn't believe that you could grow new brain cells during your adult life. This was based off the fact that nerve cells, unlike liver cells or bone cells, don't split and reproduce like most cells do. Nerve cells, however, can and do regenerate by using stem cells.

- **Giles Brindley, a doctor working to cure erectile dysfunction, dropped his pants onstage at the Urodynamics Society meeting in Las Vegas in 1983 to reveal his erect penis and prove that his experimental treatment was a success.**

- Women's health activist and Planned Parenthood founder Margaret Sanger was jailed for 1 month in 1917 for opening a family planning and birth control clinic in Brooklyn, New York.

- Green tea contains catechins, which have been shown to stabilize blood sugar levels and curb appetite.

- Jonas Salk, creator of the polio vaccine, refused to patent his discovery because he wanted it to be widely accessible to all. When asked about the patent, Salk said, "There is no patent. Could you patent the sun?"

- Laughing boosts the immune system, burns calories, and reduces stress hormones, making it a very healthy activity.

The photic sneeze reflex (also known as photoptarmosis, autosomal dominant compelling helio-ophthalmic outburst syndrome, or ACHOO) afflicts 18–35% of the population. When exposed to bright lights, these people are more likely to sneeze. Aristotle was the first to document this effect in 350 B.C.

Heart disease is responsible for 25% of all U.S. mortality, making it the number 1 killer. Cancer is a close second at 23%.

Life expectancy for a child born in the United States in the year 2007 was 77.9 years.

Women tend to have a more acute sense of smell than men. This is true at birth and remains so throughout life.

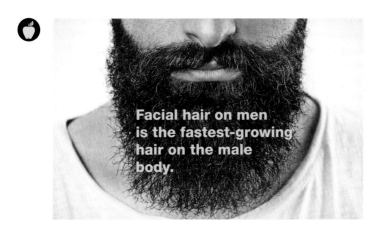

Facial hair on men is the fastest-growing hair on the male body.

Overeating has been shown to temporarily dull your sense of hearing.

Men get hiccups more often than women.

The jawbone is the hardest bone in the human body.

Women's rights activists in Germany have started sending drones into neighboring Poland loaded with pills that can safely induce abortion. The medicine is legal in Germany, but the Catholic-leaning country of Poland has strict legal limits on abortion.

Women with severe PMS are 40% more likely to develop high blood pressure later in life.

Montel Williams had a double mastectomy when he was a young man in the marines after doctors found a lump in his chest that ultimately ended up being benign.

Scientists are able to predict which antismoking ads will be the most effective by looking at the brain scans of smokers while they watch the ads.

Though chain-smokers are a common presence at AA meetings, alcoholics who quit smoking when they quit drinking are half as likely to relapse.

Cervical cancer survival rates have increased because of the Affordable Care Act (a.k.a. Obamacare). Early diagnosis is critical to beating cervical cancer, and more women have been catching their cancer early due to increased screening.

The proliferation of flat-screen TVs has increased the number of head and neck injuries in infants and children. The slimmer gadgets topple more easily.

Supermodel Christy Turlington was diagnosed with emphysema at age 31. She had started a decade-long pack-a-day habit at just 16.

The Repository for Germinal Choice was a sperm bank that accepted donations only from Nobel Prize winners or potential "future Nobel laureates." The criteria for women donating their eggs were less strict, as long as they were not gay. Only 1 Nobel laureate actually ever donated, and its founder never received a Nobel Prize but was the first winner of the Ig Nobel Prize.

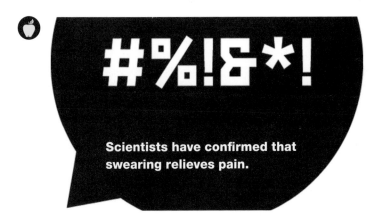

#%!&*!

Scientists have confirmed that swearing relieves pain.

Doctors suggest that sniffing twice can prevent a fainting spell.

Riding roller coasters relieves asthma symptoms.

A man found a kidney donor for his wife by posting on Craigslist, the classified advertisements website. The donor said he saw the ad by accident while browsing for construction supplies.

French physician René Laennec invented the stethoscope in 1816 because he had a patient who was so overweight, he could not hear her heartbeat.

In 1994, U.S. surgeon general Joycelyn Elders was fired because she advocated for the legalization of drugs and stated that masturbation was a natural part of sexuality.

Louis Pasteur, the French microbiologist who invented pasteurization, among other major advances, was working on the vaccine for rabies in the 1880s. During that time, he kept a loaded pistol in his lab and ordered that if he or any of his employees were to contract the virus, they should be shot in the head immediately.

In 2005, a glitch in the popular MMORPG *World of Warcraft* caused a virtual plague that killed off characters in the game. The event caught the attention of both epidemiologists, who studied the event as a model for how society might react to a similar event in real life, as well as terrorism experts, who were intrigued by speculation that the plague could have been caused by a group of nefarious hackers.

Myrtle Corbin, a 19th-century Tennessee woman, was born a dipygus, meaning she had two of everything from the waist down. She lived to be 60 years old and gave birth to 5 kids.

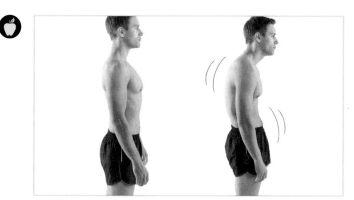

Maintaining good posture can burn up to 350 calories per day.

The pain caused by passing a kidney stone is said to be worse than giving birth or being shot with a bullet.

In 2005, a 13-year-old baby was born. It had been cryogenically frozen as an embryo, sold by a fraudulent fertility clinic, tracked down, and then finally placed back inside the mother.

Chronic diarrhea from various diseases and infections was the leading cause of death among both Confederate and Union soldiers in the American Civil War, and there was a code of honor among all soldiers to never shoot someone when he was defecating.

- Operation INFEKTION was a KGB disinformation campaign to trick people in the third world into believing that AIDS was created and spread by the United States.

- Heart cancer is extremely rare. This is because heart cells stop dividing early on in life, and cancer is characterized by out-of-control cell division.

- The oldest person in history, Jeanne Calment, smoked cigarettes from the age of 21 until she was 117. She quit 5 years before her death at age 122.

- Defibrillators are used to correct the heart's natural pulse, not for resuscitation, as is often seen in movies and on television.

- People who've undergone heterotopic procedures carry around 2 hearts inside them, 1 dead and 1 living.

- According to a study at Rutgers University, oral sex and swallowing sperm can lower the risk of preeclampsia, the dangerously high blood pressure that sometimes accompanies pregnancy.

- Movement of the jaw from talking or chewing keeps earwax from building up.

- Doctors cured blindness in 60-year-old Sharon Thornton by implanting part of one of her canine teeth in her eye. The tooth served as an anchor for a special lens that allowed her to see again.

The state of North Carolina sterilized about 7,600 people, typically against their will, between 1929 and 1974, in an effort to reduce rates of mental illness and social deviance. Among those sterilized, 2,000 were children, and in at least 2 cases, it was because they masturbated.

An obese 27-year-old man lost 275 pounds during a 382-day physician-monitored fast and suffered no medical complications.

Some doctors believe that eating your own nasal mucus might be beneficial to your immune system.

Coma patients often feel compelled to masturbate after they wake up.

Radioactive apatite is typically used to fertilize tobacco, and smoking a pack a day for a year is the radioactive equivalent to receiving 2,000 chest X-rays in the same amount of time.

The U.S. government spends a higher percentage of its revenue on health care than several countries with socialized single-payer systems.

In 1997, 685 Japanese children were sent to hospitals after a particularly intense episode of *Pokémon* triggered mass vomiting, dizziness, and seizures.

- Krista and Tatiana Simms are conjoined twins, connected at the thalamus, who can hear each other's thoughts and see through each other's eyes.

- Tea made from banana peels contains the precursor to serotonin and can help promote sleep, reduce depression, and decrease cravings for unhealthy foods.

- Calorie restriction has been shown to significantly slow aging in mice and rats.

- Lactose intolerance varies considerably by ethnicity and geographic region. For example, 90–100% of East Asians are lactose intolerant, while only 5–15% of British people are.

- Tylenol overdose is the leading cause of acute liver failure in the United States.

- If your intestines become completely clogged, it is possible that you'll vomit your own feces.

- The Salt Institute is a nonprofit-industry trade group that champions the many benefits of salt to the environment and individuals, because everything is a little better with some salt.

- An adult human heart valve is about the size of a half-dollar.

- Worldwide, 9 out of the 10 heaviest body weights ever recorded were from the United States.

In 2010, Kansas State professor of nutrition Mark Haub went on a 10-week Twinkie diet in order to demonstrate that weight loss is caused by calorie reduction and not by nutrition. He lost 27 pounds.

In Oregon, an adult patient who has been diagnosed with a terminal illness and 6 months to live may obtain a prescription for a lethal dose of medication from their physician for the purpose of committing suicide.

In 1518, there was a dancing plague in Alsace, France, in which 400 people continuously danced for 4–6 days until they collapsed. Many died from heart attack, stroke, or exhaustion.

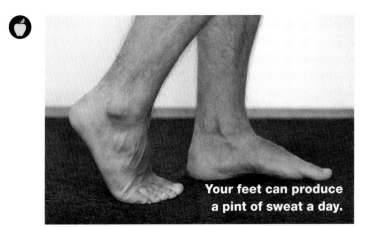

Your feet can produce a pint of sweat a day.

If you were locked in a sealed-off room, you would die of carbon dioxide poisoning before your brain shut down from oxygen deprivation.

The ringing in one's ears is called tinnitus. Often caused by loud noises, in a large percentage of the population it never goes away. It can be severely irritating, with 45% of constant sufferers reporting anxiety and depression.

Giggle incontinence is the involuntary release of urine in response to giggling or laughing. Scientists believe it is a form of cataplexy, in which muscle tone is reduced in reflex to strong emotions.

A little more than 2 years after overcoming a 12-day-long coma, 21 days on a respirator, and 2 months in a wheelchair with her femur broken in 12 places, Madeline Mitchell became Miss Alabama.

When parents instruct their children to clean their plates, they could be incidentally teaching them to ignore natural hunger cues that might cause their children to develop obesity problems later in life.

In 1962, an epidemic of uncontrollable laughter broke out in what is now Tanzania. It began among a group of schoolgirls, but spread to neighboring villages. It lasted for 6 months. Schools were closed and 1,000 people were affected in total.

In 1980, a Las Vegas hospital suspended employees who were running a betting pool on when terminally ill patients would die.

Certain regions of women's brains grow larger in the months immediately after giving birth.

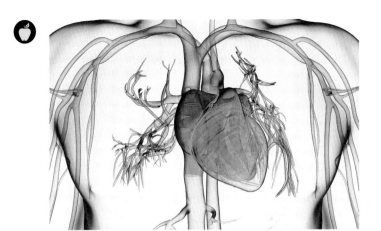

The human heart pumps 1.5 gallons of blood per minute.

Men who masturbate regularly are less likely to develop prostate cancer than men who do not. There were 2 landmark studies in 2003 and 2004, which found that middle-age men who recollected having at least 4 orgasms per week throughout their 20s, 30s, and 40s were 25% less likely to have prostate cancer than men who'd had fewer orgasms during the same period.

Anton-Babinski syndrome is a rare condition in which blind people are not aware that they are blind and continually insist that they can see, or that it must be the poor lighting, or any number of excuses short of admitting their blindness.

Drinking coffee is good for those suffering an asthma attack, because caffeine acts as a bronchodilator, opening up air passages.

Orthorexia nervosa is an eating disorder where people are obsessed with eating healthily. Those who suffer from the disorder often believe commercial food has too many harmful pesticides or bad calories and end up starving themselves.

Mite colonies are found on the eyelashes of 80% of people 60 and older.

Crying makes 9 out of 10 people feel better and healthier. The hormones and minerals released by emotional tears help regulate many body functions. Researchers have suggested induced weeping as a form of therapy.

Britain taxes women's menstrual products as a luxury item. Meanwhile, men's razors and crocodile meat are not considered luxe and are not taxed.

Researchers believe a drug used to treat diabetes may allow people to live much longer lives, perhaps allowing many people to survive into their 120s.

The first pacemakers had to be plugged into the wall. Recently, scientists have developed a thin electrified sheet that makes the heart beat on its own.

Inspired by the common legend that German monks fasted on beer during Lent, one beer geek decided to see if he could survive drinking only beer for 46 days. He did and said it left him feeling rejuvenated and 25 pounds lighter.

The Hum is an acoustic phenomenon that occurs all over the world in very specific places that only a small fraction of the population can hear. It's a low-frequency sound, found at between 32 Hz and 80 Hz, and has been linked to at least 3 suicides in the United Kingdom.

A landmark study has found few structural differences between the brains of men and women, contradicting previous understandings of the brain. While there are parts of the brain that are considered typically male or typically female, most people have a mosaic of male and female characteristics in their brains.

Gut bacteria may cause rheumatoid arthritis. The bacteria may be training the immune system to attack joints. They have also been linked to obesity, allergies, and depression.

Many doctors are abandoning the traditional white lab coat. The long sleeves are easily contaminated, and studies have shown that doctors rarely wash them. However, many doctors are reluctant to give them up, saying they inspire trust.

Popular energy drinks have been shown to increase the risk of heart disease.

Life expectancy for almost every demographic group has been rising for the past decade, except for middle-age white Americans. Increased suicide and drug abuse rates are thought to be the cause.

An analysis of scientific papers found that many scientists may not be reading the papers they cite: 2.4% of citations were critical.

Women's testosterone levels rise when they are in positions of power, even when they are just acting or pretending to have power.

The average stomach holds 3 quarts of food and drink. When it reaches its full limit, about 5 quarts, it will burst.

Insulin has been shown to increase dopamine levels by up to 55%, which helps explain why eating sugar makes you happy. Some evidence has linked insulin resistance with depression.

You excrete an average of 3–8 ounces of feces a day.

A 15-year French study found that bras are useless. In fact, wearing a bra may actually cause breasts to sag.

1. **Riding roller coasters can be good for relieving what?**

a. Headaches

c. Asthma symptoms

b. A stuffy nose

d. Stress

2. **Overeating has been shown to temporarily dull your ability to do what?**

a. Hear

c. Smell

b. Make decisions

d. See

3. **The average person farts how many times a day?**

a. 25

c. 14

b. 5

d. 100

4. **Putting what on an open wound will help stop the flow of blood?**

a. Vaseline

c. Salt

b. Raw sugar

d. Water

Psychology

- Sarcasm makes you more creative.

- Staying up late leads to more nightmares than going to bed early.

- Men are more likely to dream about sex than women. Women tend to have more nightmares than men.

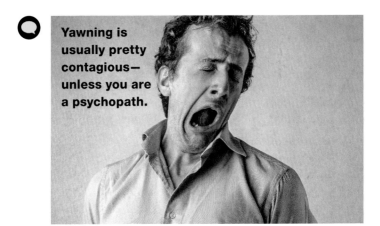

Yawning is usually pretty contagious— unless you are a psychopath.

- People who regularly play video games are more likely to be able to control their dreams while they're dreaming.

- People who live in areas with nice weather and beautiful scenery tend to be less religious.

- White Christians no longer make up a majority of the American population, just 46%.

- The most commonly spoken language after English and Spanish in 16 states is German; in Michigan it's Arabic; in Illinois it's Polish; and in South Dakota it's actually Dakota.

- Members of the Walton family, who own Walmart, are the wealthiest individuals in 3 different states (Texas, Arkansas, and Wyoming). The family has as much money as 42% of the least wealthy Americans combined.

- North Dakota is the only state in the country where the wealthiest person isn't a billionaire.

- Research has shown that using the word "fancy" to describe items on menus increases sales by 27%.

- American 18- to 34-year-olds spend 25.7 hours on Facebook, 7 hours on Instagram, 5.9 hours on Snapchat, 5.7 hours on Tumblr, and 3.5 hours on Twitter each month.

- Rich people are more likely to drive through pedestrian crosswalks and cut off other cars.

- People commit suicide more often on Mondays.

- The United Nations says that 95% of online abuse is directed toward women.

- Researchers have tracked how the smile has evolved in yearbook portraits. Around the turn of the 20th century, most students had a neutral expression (easier to hold for the long exposures required in photography's infancy). Women started smiling in the 1960s, but big smiles from men didn't come until the 1990s.

- Finland, Norway, and Sweden rank as the best countries for workingwomen, according to economists. Poland comes in at number 4 after the Scandinavian countries. The United States comes in at only number 17.

-

People cry most often in the evening. The most common time people cry is between 6:00 and 8:00 P.M.

- The Gruen transfer is the feeling people have when they enter shopping malls and become disoriented by the intentionally confusing layout, forgetting what they intended to buy.

- 36% of Japanese men have absolutely no interest in having sex.

- Loneliness poses as big a health risk as obesity. Studies have found that loneliness increases your mortality risk by 26%.

- Men eat more food in front of women than they do in front of other men. Women generally eat the same amount with men and women, but usually feel as if they've overeaten or were rushed through their meal when they eat with men.

- Studies have found that people who tend to stay up late are smarter, more prolific lovers, and even better at baseball. However, night owls are more prone to bad habits than people who normally wake up early and are not as happy.

- Those who score higher on IQ tests are more likely to trust others than those on the lower end. One explanation is that more intelligent people are better at judging character and form relationships with people who are less likely to betray them.

- According to data from several U.S. Censuses, Illinois is the most average state in the union.

- Married people are happier than unmarried people.

- Mathematicians have concluded that it is statistically impossible for vampires to exist because they would decimate the human population within 3 years.

- Many introverts are covert narcissists.

- People are more likely to agree with a statement written in Baskerville than any other font.

- Instances of sexual abuse against horses are on the rise in Switzerland. About 10,000 Swiss citizens are thought to be predisposed to zoophilia—a sexual attraction to animals.

- The most isolated inhabited island is the 8-mile-wide Tristan da Cunha, 1,743 miles away from Cape Town, South Africa. The British territory has a population of less than 300, most of whom are mostly farmers. There is no airport, electricity is supplied by generators, and there is only one 1 road. The island is in fact a volcano that last erupted in 1961.

- Suicide rates in Japan spike in March, the end of the fiscal year.

- American civilians have the highest rate of gun ownership per capita than anyone else in the world, even beating out war-torn countries such as Syria and Iraq. In fact, there are more guns in the United States than people.

- Stores are starting to use facial recognition technology to identify shoplifters.

- America has the world's most anxious population. Studies have shown that 18% of the population has an anxiety disorder, making it the most common mental illness. Women are generally more affected than men.

- The average Facebook user spends a dollar a day on data just using Facebook. Meanwhile, 20% of the world's population lives on less than a dollar a day.

- The leading cause of disability in the United States for adults under the age of 45 is depression.

- Agoraphobia is a panic disorder and social anxiety where people become severely anxious in crowded public places. The term comes from *agora,* the Greek word for "market."

- Spain and the United States are tied for highest rates of cocaine consumption. Fortunately, cocaine use has declined significantly in the last decade.

- Aokigahara, a dense forest in Japan, is infamous as a suicide site. In 2003, 105 bodies were found, most deceased as a result of hanging.

● Republicans prefer politicians with deeper voices.

● One of the largest cities in the world is occupied for only about 5 days. Each year during the Hajj to Mecca, 3 million pilgrims stay in the tent city of Mina, Saudia Arabia. After the Hajj is over, the city is practically abandoned.

● People aged 41–70 are more likely to use e-mail to communicate at work than so-called digital natives. Those aged 21–40 report that they prefer face-to-face communication in the workplace.

● Teenagers are happier than ever, while adults in their 30s have become increasingly unhappy since the 1970s. Scientists blame both trends on the Internet.

●

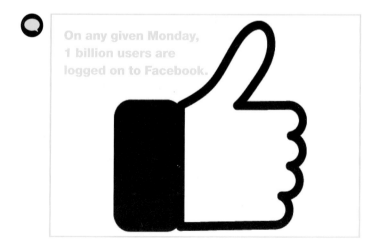

On any given Monday, 1 billion users are logged on to Facebook.

- Ending e-mails with a statement instead of a question is more efficient and leads to fewer e-mails.

- A correlation survey by two economists suggests that money can actually buy happiness.

- Apparently people can solve anagrams faster while lying on their backs.

- Scientists found that the most relaxing song ever is "Weightless" by the Marconi Union—it reduced anxiety by 65% in the average test subject.

- Researchers at the University of Chicago determined there are 120 varieties of knots that form spontaneously on the cords of headphones when left inside the pocket.

- Humans tend to subconsciously mimic people with heavy accents out of empathy.

- Cryptomnesia is the phenomenon of thinking you invented a joke or idea, when in fact you heard it somewhere else previously.

- In 1964, researcher Milton Rokeach put 3 schizophrenic men together who all believed they were Jesus Christ to see what would happen. After the study, Milton said that while the subjects were not cured of their delusions, "it did cure me of my godlike delusion that I could manipulate them out of their beliefs."

- Showing people facts that contradict their beliefs tends to actually reinforce their incorrect beliefs.

- 74% of humans have reported experiencing derealization, or when reality feels dreamlike.

- Princess Alexandra of Bavaria had a mental disorder that caused her to believe she had swallowed a piano made out of glass.

- People suffering from body integrity identity disorder want to have a limb or limbs amputated, and they often express envy for amputees.

- "Princess sickness" is a term used in China and Korea to describe a psychological phenomenon affecting young women that is characterized by narcissism, egocentrism, and melodramatic personality resulting in individuals acting like princesses.

- In a study where Monopoly players were given twice as much money as their opponents, the "rigged" players displayed cruelty toward their opponents.

- In the United States, 450 people die each year from falling off their beds.

- Li is the most common surname in Canada.

- Sri Lanka is the country that looks up the word "sex" the most on Google.

You blink over 10 million times a year.

The average 4-year-old laughs 300 times per day, whereas the average 40-year-old laughs 4 times per day.

On average, the United States consumes 100 acres of pizza every day.

Over 50% of the U.S. population believes in psychic phenomena.

Hermann Rorschach, famous for pioneering the Rorschach inkblot test, was actually nicknamed "Inkblot" as a child.

According to UN estimates, October 12, 1999, was the day the earth exceeded 6 billion residents.

As of 2014, the country with the highest life expectancy is Monaco at 89.57 years. Chad has the lowest life expectancy at 49.44 years.

Since 1973, 156 people have been exonerated and freed from death row. Florida alone has freed 26 death row inmates in the past 4 decades.

The Mars Curiosity rover project cost $2.5 billion, which is less than 0.5% of U.S. military spending in 2015.

At any given time there are roughly 2 million impaired drivers on the road in the United States.

You have a 1 in 2,067,000 chance of dying in a plane crash.

Thinking about sex releases hormones that help you concentrate.

You have a 1 in 423,548 chance of dying by falling out of your bed.

Terrified of making small talk at dinner? That's called deipnophobia.

Cathisophobia is the fear of sitting down.

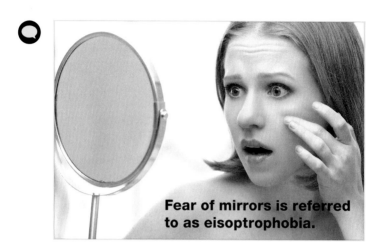

Fear of mirrors is referred to as eisoptrophobia.

Nyctohylophobia is a fear of the woods at night.

Over 50% of the 1.1 billion people living in Africa are under the age of 25.

To be in the top 1% of earners in Connecticut, you would need to make $678,000 a year or more. However, if your crossed the border into Rhode Island, you would have to make only $315,000 to be in the 1% club.

Arabic, English, and Swahili are the 3 most popular languages spoken on the African continent, though there are over 2,000 recognized languages spoken in Africa. Africa is the most language-rich continent on the planet.

There are more people in New York City with Internet access than in all of Africa.

Americans spend more on jewelry, shoes, and watches ($100 billion) than on higher education ($99 billion).

Over 30% of Americans buy holiday presents for their pets.

British researchers found that the average 10-year-old owns 238 toys but plays with just 12 daily.

The average person will spend a total of 3,680 hours, or 153 days, searching for misplaced items. Keys, cell phones, sunglasses, and paperwork top the list of commonly lost items.

Facebook users spent $1.85 billion on virtual purchases, from *FarmVille* chickens to virtual presents.

There are roughly 7 billion people in the world. Roughly 4 billion own a mobile phone, while only 3.5 billion own a toothbrush. That means there are more cell phone owners than toothbrush owners!

About 750,000 people, roughly 2.5% of the population of the United States, perished during the Civil War.

Over half of professional pilots have admitted to falling asleep while operating an aircraft.

Every second an hour of video is uploaded to YouTube.

There are enough credit cards in circulation to wrap around Earth's equator 3 times.

Credit cards expire because the magnetic strip demagnetizes over time.

Credit Card card numbers are formulated using a Luhn algorithm. To see it in action, double every second number on your credit card and add them up. If the number is divisible by 10 you have a real credit card (or a very clever forgery).

The average household receives an average of 6 credit card offers a month.

VISA is a recursive acronym for Visa International Service Association.

Every second there are over 10,000 credit transactions around the world.

In 2013, the average American had $5,000 in credit card debt and 1.96 credit cards.

The first digit of the a credit card represents the type of card: 1 and 2 are for airlines; 3 is for entertainment and travel; 4, 5, and 6 are for banking; and 7, 8, and 9 are for other uses.

- The average rural African has to walk 3.7 miles for potable water.

- Between bathing, drinking, going to the bathroom, and the food we eat, the average American uses over 100 gallons of water a day. That's enough to fill a small hot tub. The average resident of sub-Saharan Africa uses about 3 gallons of water a day.

- Chicago has the largest population of Polish citizens outside of Poland.

- **According to one study, only 38% of men and 60% of women wash their hands after they use the bathroom.**

- People with dilated pupils are rated as more attractive. This is because our pupils dilate when something interests us. If you see someone with dilated pupils, you assume they are interested in you and you find them more attractive.

- A "Duchenne smile" is the scientific term for a real smile of genuine happiness as opposed to a forced or polite smile.

- One study found that waitresses who lightly touch a patron when delivering the bill got a higher tip. This is thought to be because the hormone oxytocin, which is associated with love and bonding, is released when humans touch.

- About 75% of all U.S. Internet searches occur through Google.

- Google processed 11.382 billion searches in September 2015.

- 56% of Internet users have googled themselves.

- Men have an average of 7 sexual partners per lifetime while woman have an average of 4 sexual partners.

- During sex, 75% of men and 29% of women report always reaching orgasm. This is popularly referred to as the "orgasm gap."

- 10% of married adults report sleeping alone most nights.

1 in 8 Americans has worked for McDonald's at some point.

● The average male will lose his virginity at age 16.9, the average female at age 17.4.

● Most men have about 3–5 erections during an 8-hour night of sleep.

● The ability to delay gratification as a child has been linked to success later in life. A 1962 Stanford study asked 4-year-old children if they would rather have 1 marshmallow now or 2 later. Most chose to wait but couldn't last the required 15 minutes with a plate of marshmallows in front of them. Those who did last the full 15 minutes typically used strategies like closing their eyes or not looking at the marshmallows. Over the course of their lives, the children who were able to wait the full 15 minutes had lower rates of obesity and addiction and achieved greater levels of professional success.

● Academy Award–winning actors and directors tend to live longer than their peers who were nominated but lost.

● The "bystander effect" is the term psychologists use to describe the phenomenon that people who are in distress often have trouble getting help in crowded areas. Onlookers tend to think that someone else will step in and help. People in distress often have an easier time getting help when there are fewer people around. To break through this effect, researchers often suggest singling out 1 person in the crowd, making eye contact, and demanding that they specifically help you.

Children who see an adult hitting or acting aggressively toward a doll are more like to do the same thing when allowed to play with that doll than children who observe adults acting passively toward a doll.

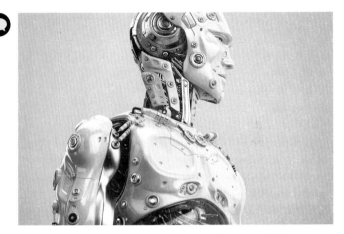

20% of UK residents admitted they would be open to having sex with a robot.

Yawning is not only contagious among people. Dogs who watch people yawn are significantly more likely to yawn themselves.

Researchers once put fake missing children flyers in a store. Some people looked the posters over while others just glanced in passing. Not a single person noticed that the child in the posters was in fact in the store.

Indonesian pirates are typically less violent than Somali pirates.

- 50% of Americans believe themselves to be more intelligent than the average American. White men are more likely to believe themselves to be smarter.

- Since 9/11, white right-wing American terrorists have killed twice as many people in the United States as Muslim extremists. Of the 26 terrorist attacks on American soil since 9/11, 19 of them were carried out by white supremacists or radical antigovernment extremists.

- Psychopathy appears in children as young as 2 years old. Neuroscientists aren't quite sure what the causes are.

27% of Americans believe that God decides the outcome of sporting events.

- Couples who have sex once a week are the happiest. Studies have shown that having sex more often than that did not necessarily lead to a more satisfying relationship.

- Researchers have started tracking the spread of infectious diseases by looking at cell phone metadata.

- **Toddlers pick up fat-shaming attitudes from their mothers as young as 11 months.**

- Twitter can predict when and where heart attacks are most likely to occur based off language analysis.

- Weather has caused 90% of the world's disasters in the last 20 years.

- Children perform better in school if they stand at their desks instead of sit. Standing in school can also help prevent childhood obesity, as it burns 15–25% more calories.

- Studies suggest that online therapy to treat depression is far less effective than human contact, and that online therapy methods, when used alone, showed little improvement in symptoms.

- Research has shown that exercise is just as effective in treating depression as antidepressants are.

- A neuroscientist was studying the brain scans of murderous psychopaths as a side project and discovered that he himself was a psychopath. Though, as far as we know, he never murdered anyone.

Antidepressants are the 3rd-most prescribed medication, after painkillers and cholesterol reducers.

4 out of 10 workers believe that having a flexible work schedule will improve their sex lives.

Most mental illnesses show themselves in late adolescence. That's when the prefrontal cortex—the most emotionally complex part of the brain—begins to mature.

The stresses of living in poverty are equivalent to losing approximately 13 IQ points.

A Chinese woman who went missing at age 14 was found 10 years later living in an Internet café. Authorities fined her $156 and convinced her to reunite with her parents.

Looking into special light boxes for 30 minutes each morning has been shown to be an effective treatment for depression.

- Lap dancers earn more when they're ovulating.

- Prawo Jadzy was thought to be Ireland's most reckless driver, racking up dozens of speeding and parking tickets. It turns out Prawo Jadzy is Polish for "driver's license."

- A customer satisfaction report ranked only 3 American airports in the top 50, with Cincinnati coming in highest, at number 30.

- It's been mathematically proven that organizations are most efficient when people are promoted randomly.

- People think they are more attractive when they are drunk.

- Psychologists have determined that leaning to the left makes the Eiffel Tower look smaller.

- ## Our favorite songs tend to be associated with a personal emotional event.

- Spending money on experiences rather than material things tends to make people happier.

- Research suggests that associating with happy people tends to enhance personal happiness.

- The most stressed members of the population are 18- to 33-year-olds.

After a study found that over half of women in Tokyo, Japan, had reported being groped inappropriately on the train, the government introduced female-only train cars during rush hour.

The Dunning-Kruger effect describes how smart people tend to doubt themselves, while less intelligent people tend to be overconfident.

When you recall a past event, you're actually remembering the last time you remembered it.

43% of Americans say they have assigned seats in their living rooms.

Thinking in a foreign language tends to reduce bias.

Announcing your plans makes you less motivated to complete them.

Neurotic people cry more than others.

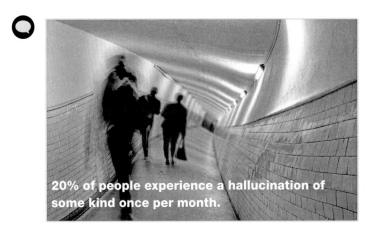

20% of people experience a hallucination of some kind once per month.

In a 2011 survey of 15- to 30-year-olds in Afghanistan, 92% had never heard about the events of 9/11.

Students in a British study tended to enjoy a task less the more they were paid for it.

Some studies suggest that 80% of the power of Prozac is placebo.

Fantasizing about future success actually reduces motivation.

Group brainstorming tends to engender intellectual laziness and reduce the overall creativity of results.

A 1987 study found that thought suppression is extremely counterproductive, even in the short term.

● A 1981 study suggests major life events are not as important to our mental health and overall success as our day-to-day lives.

● Researchers found that when water was dropped into the tear ducts of both eyes to simulate tears, 54% of test subjects reported sadness, but 29% said they felt cheerful.

●

A survey of librarians found that 3% would like to marry Cher.

● In a recent survey about technology terms, 11% of respondents believed "HTML" to be a sexually transmitted disease, 27% thought a "gigabyte" is a South American insect, 12% identified "USB" as an acronym for a European country, and 15% believe that "software" refers to comfortable clothes.

- People who violate social norms tend to be seen as powerful by others.

- Relaxed shoppers tend to spend more money than stressed ones.

- Canadians, on average, have larger penises than Americans. But Americans have bigger breasts than Canadians. Russian women have the biggest breasts in the world, generally.

- A study at Cornell University found that groups with more than 1 narcissist tended to have the most creative brainstorming sessions due to the competition between the narcissists to be the center of attention.

- A Columbia University study suggests that having many choices, contrary to popular belief, can actually sharpen focus on finding the best qualities.

- Anger can actually reduce confirmation bias.

- The chemical brain profile of people experiencing romantic love is virtually indistinguishable from people with obsessive-compulsive disorder.

- The average woman speaks 20,000 words per day—13,000 more than the average man.

- Paris syndrome is a severe form of culture shock experienced most frequently by Japanese tourists who become depersonalized and delusional after their encounter with the French capital fails to live up to expectations.

- The average American high school kid today has the same level of anxiety as the average American psychiatric patient in the 1950s.

- There has never been a case of a congenitally blind person developing schizophrenia.

- Anorexia has the highest mortality rate of any psychiatric disorder. Up to 20% of those who suffer will eventually die from causes related to the disease.

- "Hikikomori," or acute withdrawal syndrome, is a term to describe the 1 million or so Japanese men who intentionally isolate themselves from society, often due to a sense of perceived failure.

- In the past 2 decades, $26 billion in ransom has been paid out in the United States.

- The odds of being killed by a dog, worldwide, are 1 in 147,717.

- 9% of U.S. citizens don't carry cash, and half carry $20 or less.

There are over a billion cars on the world's roads.

Couples with silk sheets have sex an average of 4.25 times a week, twice as much as couples with cotton sheets.

A male researcher found that his success rate at getting a woman's phone number was around 9%, but jumped to 28% when he had a dog with him.

80% of sexually active adults will contract an STD at some point in their lives.

Every 30 seconds, 106,000 aluminum cans are used in the United States.

Having children makes people less empathetic and more suspicious of others.

Every year, 10 people are killed in the world by falling vending machines.

The average car requires 39,000 gallons of water to make.

30% of people will not sit on public toilets.

According to a survey, a typical British thirtysomething lives with their partner, drives a hatchback, and loves *Dirty Dancing,* and their favorite band is Queen.

90% of 18- to 24-year-old respondents said they would trust medical information shared on social media networks.

Researchers at Cambridge University found that happy children are not more likely to get married than unhappy children when they grow up, and they are actually more likely to end up divorced.

Men with purple bedrooms have twice as much sex as those with blue or gray bedrooms.

The world's deepest mailbox is 10 meters underwater in Japan.

Parents have started naming their children after Instagram filters. The most popular filter name was Lux, but there were even a few Kelvins.

Baker-Miller Pink is a shade of pink that is said to be calming and reduce aggression. The bubble gum color has been used in jail cells and even worn by boxers to throw off their opponents.

34% of American men describe themselves as "totally normal."

25% of American men are 6 feet or taller.

○ Only 30% of people can flare their nostrils.

○ The German word for "loser," someone who is hopeless and pathetic, is *gurkentruppe,* which translates literally to "troop of cucumbers."

○ Only 6% of coupons are redeemed.

○

The average American spends 90 minutes a day looking at their cell phone, which adds up to a full 23 days each year, or nearly 4 whole years over a lifetime.

○ The halo effect is a cognitive bias in which our impression of a person (e.g., they were kind) influences our judgment of either a single trait in that person or the person's overall character (e.g., they are good).

- A 2013 study from the University of Virginia found scientific evidence of how significantly empathy impacts us. Essentially, when a person's brain witnesses powerful events being experienced by others, it perceives them as if they're happening to the person.

- When asked to judge the behavior of someone who knowingly had sex with someone while diagnosed with a treatable, nonfatal sexually transmitted disease versus someone who had sex while diagnosed with swine flu—which has a high fatality rate—most people were more likely to see the behavior of the person with the STD as risky and immoral.

- In 2011, prominent Dutch psychologist Diederik Stapel was caught fabricating the results of a research study. The revelation prompted reviews of Stapel's earlier work, which ultimately led to the redaction of 50 landmark studies the scientist had overseen.

- 68% of people suffer from phantom vibration syndrome, or believing one's phone is vibrating when it is not.

- Solomon Asch's groundbreaking psychological study on conformity posed a multiple-choice question to a group of 9 men. The thing is, 8 of those men were told which answer to give in advance. When a higher percentage of the confederates were told to give the correct answer, the subject typically gave the right answer. When they were told to give the wrong answer, so did the subject in most cases—50% of people gave the same wrong answer as the others in over half the trials.

- Talking to yourself makes your brain work more efficiently.

- Neuroscientists have shown that people love their dogs as much as their children.

- Women cry an average of 47 times a year while men cry just 7.

- An Israeli study found that when asked about their feelings about immigrants, women who had children present with them when questioned tended to have more negative views.

- Your brain "rewrites" boring speeches or lectures to make them more interesting.

- Children who appear distracted or have wandering minds have a better working memory and recall.

More than a quarter of all Americans believe that the sun orbits Earth.

The "Walker Effect" is the name of the brief sense of imbalance people often feel when stepping onto an escalator that is not moving, even when they are fully aware the escalator is broken.

77% of people in heterosexual relationships believe that men should pay for the first date, a figure that has not changed much in the past 30 years. 39% of women admitted that they hoped men would reject their offers to pay or split the bill.

People are more likely to help a stranger if the stranger has a dog with them.

New mothers spend an average of 14 hours a day thinking about their newborns.

76% of women and 74% of men believe that the first American female president will be a brunette.

● Adopted children are significantly more likely to have learning disabilities and behavioral problems than children raised by biological parents.

● Methamphetamine was prescribed to treat depression in the 1930s and 1940s.

● People tend to agree more often with statements such as "Some people deserve to be treated like animals" after spending a lot of time with their family, especially after Thanksgiving.

● 95% of Americans have sex before marriage.

● A poll found that more people are likely to support the Affordable Care Act than Obamacare, even though they are same thing.

● People who spend 2–3 hours a day on social networking sites are more likely to identify as "big spenders."

1. **People commit suicide more often on what day of the week?**

   a. Friday

   b. Tuesday

   c. Thursday

   d. Monday

2. **How much time total does the average person spend looking for misplaced items?**

   a. 18 hours

   b. 153 days

   c. 47 weeks

   d. 1 year

3. **Thinking about what releases hormones that can help you concentrate?**

   a. Food

   b. Music

   c. Sleep

   d. Sex

4. **Research has shown that doing what is just as effective at relieving depression as antidepressants are?**

   a. Listening to music

   b. Reading a book

   c. Exercising

   d. Playing video games

Pop Culture

Professional wrestler Big Van Vader was once interviewed on a TV show in Kuwait. He was told to attack the interviewer if asked if pro wrestling was fake. The interviewer was not in on the gag, and Vader was charged $164.

During the filming of *The Godfather,* prankster Marlon Brando put rocks under the bed in the scene where Don Corleone is being carried upstairs.

Mariah Carey is a beauty school dropout.

Dustin Carter, from Hillsboro, Ohio, was an accomplished high school wrestler without arms or legs.

At age 24, Evan Spiegel, the founder of Snapchat, is the world's youngest billionaire.

Robert Plant of Led Zeppelin hates the band's biggest hit, "Stairway to Heaven," so much that he once donated money to a public radio station that announced it would never play the song again.

To date, Psy's "Gangnam Style" is the most viewed video ever on YouTube, with nearly 2.5 billion views.

Disneyland banned men with long hair up until the 1960s. It still does not allow its employees to sport beards.

Sting's *Ten Summoner's Tales* was the first item ever sold on the Internet, in 1994.

We're all pronouncing Voldemort wrong. According to Harry Potter author J. K. Rowling, the "t" is silent, like the French word for "death," *mort*.

J. K. Rowling apologized to fans for killing off Fred Weasley.

The airport in Gibraltar, Spain, has a busy street that goes straight through the runway.

John Wayne's real name was Marion Morrison.

Salt Lake City, Utah, has the highest population of plastic surgeons per capita in the United States.

On the TV sitcom *Seinfeld,* Jerry Seinfeld had 66 girlfriends over the course of the show; 2 of them were named Nina. George Costanza had 47 girlfriends, 2 of whom were also named Nina (1 of the Ninas was the same woman).

The New Age singer Enya lives in a castle in Ireland called Manderley Castle, which she named after the castle in her favorite book, Daphne du Maurier's *Rebecca.* Bono is one of her neighbors.

People who live in Vermont use the smiling poop emoji more than any other state.

3 out of 4 Americans use an emoji in text messaging every single day.

The playwright Eugene O'Neill was born in a hotel room and died in one as well.

Blowing into Nintendo cartridges didn't actually make them work. In fact, the moisture in the breath corroded the metal prongs that connected the game cartridge to the console.

The billionaire, and so-called Queen of Mean, Leona Helmsley, left $12 million to her dog, Trouble. That's more than she left to her grandchildren, 2 of whom received nothing.

State Farm's jingle "Like a Good Neighbor" was written by Barry Manilow. He also wrote the "Stuck on Band-Aid Brand" jingle.

Bugs Bunny's rap in *Space Jam* was written by Jay Z.

Will Smith met Jada Pinkett Smith when she auditioned for a role as his girlfriend on *The Fresh Prince of Bel-Air*. She didn't get the part, but she did end up marrying him.

Saddam Hussein had an international corporation for laundering money called Montana Management, after *Scarface*'s Tony Montana.

Han Solo was almost killed off, but George Lucas decided to let him live because he figured a dead Solo would sell fewer toys.

In *Jurassic Park,* the *Tyrannosaurus rex*'s roars were a combination of dog, penguin, tiger, alligator, and elephant sounds, and its footsteps were from cut sequoias crashing to the ground.

The first movie ever broadcast on the Internet was *Wax, or The Discovery of Television Among the Bees,* in 1993. Back in the age of dial-up, it was shown in black-and-white, was often silent when it wasn't supposed to be, and was shown at 2 frames per second (most films are shown at 24 frames per second). An underground cult film, it is considered largely unwatchable even at Netflix speeds.

Chris Brown owns 14 Burger King fast-food locations.

After his death, Hunter S. Thompson's ashes were made into fireworks and shot from a 150-foot cannon.

The accent you hear in movies and newsreels from the 1930s and 1940s was a combination of British and American accents taught in American acting schools. Most people didn't actually talk like that and the accent quickly lost favor after World War II.

The *Brady Bunch* kids didn't have a toilet in their bathroom. At the time, television networks were forbidden from showing toilet bowls on screen. Though it was okay to show the tank part of the toilet, the producers decided it was too difficult to film.

Pillow fights are banned at the U.S. Military Academy at West Point.

The PG-13 rating was introduced after parents complained about violence in *Indiana Jones and the Temple of Doom*. The first film to be released with the rating was the Patrick Swayze–led *Red Dawn*.

A professional wrestler was once arrested for throwing a boulder through a McDonald's window for not serving him after closing time. Another was arrested for urinating on a flight attendant midflight.

Kelsey Grammer was nominated for an Emmy for his portrayal of Frasier Crane on 3 different shows: *Cheers, Frasier,* and a guest spot on *Wings*. He played the character for over 20 years.

A mannequin in a bridal shop in Chihuahua, Mexico, is thought to be an embalmed corpse or perhaps haunted. It's stood in the shop window for over 75 years, and locals are convinced that it moves at night.

Nas is responsible for many of Will Smith's raps, including "Gettin' Jiggy wit It" and "Miami."

The *Ghostbusters* firehouse is fully operational. If you catch the firefighters during downtime, they'll give you a tour.

The world's oldest socks were in fact designed to be worn with sandals. Made in Egypt sometime in the 4th or 5th century, the wool socks have two toes.

The voice of Charlie Brown's teacher was created by using a plunger mute on a trombone. A common jazz technique, trombone plungers are very similar to common toilet plungers.

Jonathan Goldsmith, the actor who portrays the most interesting man in the world in the Dos Equis commercials, lives aboard a sailboat in Marina del Rey, California.

Vin Diesel began his career after he broke into a theater with the intent to vandalize it. The director caught him, handed him a script, and asked him to read, then offered him a part in an upcoming stage production.

- Following a plane crash in 1951, Clint Eastwood and his pilot swam 3 miles to shore.

- Harrison Ford has an ant and a spider species named after him.

- Bill Murray was incredibly difficult to contact on the set of *Groundhog Day,* a movie he hated working on. When asked to hire a personal assistant, he obliged and hired a man who was mostly deaf and mute and used only sign language. Nobody else on set knew sign language, including Murray.

- Whoopi Goldberg once worked as a phone sex operator.

- While composing the score to the *Phantom of the Opera* sequel, Andrew Lloyd Webber's cat hopped onto his keyboard and deleted the entire project.

- According to producer Dr. Dre, Eminem recorded the entire *Slim Shady* LP under the influence of MDMA.

- A 1982 study found a 34% increase in shopping time in retail establishments where slow music was played.

- Jack White edits all of his recordings manually with a razor blade.

At the moment John Lennon was pronounced dead, the Beatles song "All My Loving" came on over the hospital's sound system.

In 1972, the band Deep Purple's music knocked 3 concertgoers unconscious because it was so loud.

In the United States, cities with higher rates of country music played on the radio also have higher suicide rates.

In the 1924 Soviet film *Aelita,* cosmonauts travel to Mars where they rescue a race of aliens from capitalism.

In Marfa, Texas, *No Country for Old Men* had to halt filming for a day because of a huge cloud of smoke from the set of *There Will Be Blood.*

Keanu Reeves donated $50 million to the special effects team of *The Matrix,* nearly his entire earnings from the film.

■ Vin Diesel has a twin brother named Paul who looks like Diesel's late *Fast and the Furious* costar Paul Walker.

■ Jackie Chan does not like the Rush Hour series because he does not appreciate the action scenes nor understand the American humor. He made the sequel only because they offered him an "irresistible" amount of money.

■ O. J. Simpson was supposed to play the Terminator, but James Cameron didn't want to cast him because he didn't think Simpson would be "believable as a killer."

■ Frank Zappa's album *Jazz from Hell* received a parental advisory sticker even though it is a collection of instrumental pieces containing no lyrics.

■ In 1991, Per Yngve Ohlin, better known as "Dead," the lead singer of the influential Norwegian black metal group Mayhem, committed suicide by self-inflicted gunshot wound. Dead left a suicide note that read, "Excuse all the blood, cheers." After discovering the body, Mayhem guitarist Euronymus made a necklace out of Dead's skull fragments and took a photo of the corpse, which was later used as the cover art on a Mayhem bootleg EP.

■ Bob Marley gave credit for "No Woman, No Cry" to Vincent Ford, a friend who ran a soup kitchen, to ensure the royalty checks would keep it open.

The lyrics of the number 1 singles on the *Billboard* charts over the past 10 years average at a 3rd-grade reading level.

"Channel drift" is a term used to describe when a TV channel loses its original purpose. For example, The Learning Channel's show *Here Comes Honey Boo Boo*.

In 1939, the *New York Times* predicted that the television would fail because the average American family would not have enough time to sit around watching it.

Groucho Marx's last words were supposedly "Die, my dear doctor! That's the last thing I shall do!"

Despite being the film's title character, Godzilla appears on-screen for only about 8 minutes in the 2014 *Godzilla* remake. The film's director, Gareth Edwards, said he was inspired by *Jaws:* "The film does not immediately show the beast, but rather builds up to its appearance while still delivering an eerie and terrifying off-screen presence." As an homage, the protagonist in *Godzilla* is named Brody, after the main character in *Jaws.*

As of 2016, King Kong has appeared in 7 films: *King Kong* and *The Son of King Kong* (both 1933), *King Kong vs. Godzilla* (1962), *King Kong Escapes* (1967), *King Kong* (1976), *King Kong Lives* (1986), and *King Kong* (2005). The King Kong character is scheduled to make 2 more film appearances: in 2017's *Kong: Skull Island* and in 2020's *King Kong vs. Godzilla.*

Scrabble is so popular, it inspired 2 game shows: *Scrabble,* which ran from 1984 to 1990 and was hosted by Chuck Woolery, and *Scrabble Showdown,* which ran from 2011 to 2012 and was hosted by Justin Willman.

Hattie McDaniel was the first African American woman to win an Academy Award. She received the Best Supporting Actress award for her role as Mammy in *Gone with the Wind*. McDaniel was also the first African American woman to sing on the radio.

Al Pacino boycotted the 1972 Academy Awards. He was angry about being nominated for Best Supporting Actor in *The Godfather,* as opposed to Best Lead Actor, noting that his character had more screen time than his costar Marlon Brando—who was nominated for and won the Best Lead Actor Oscar.

Spike Jonze, director of 1999's *Being John Malkovich,* claimed that when pitching the screenplay for the film, Hollywood producers urged him to instead create a film about being Tom Cruise.

The American Film Institute voted "Frankly, my dear, I don't give a damn" from 1939's *Gone with the Wind* the best movie quote of all time.

The American Film Institute voted the score for 1977's *Star Wars* the best film score of all time.

The American Film Institute voted "Over the Rainbow" from 1939's *The Wizard of Oz* the best original song for a film.

The 1946 film classic *It's a Wonderful Life* was voted the most inspiring film of all time by the American Film Institute. *To Kill a Mockingbird,* from 1962, holds the number 2 spot.

Hannibal Lecter, from *The Silence of the Lambs,* was voted the best movie villain by the American Film Institute.

Chuck Palahniuk, author of the novel *Fight Club,* felt the movie was better than the book. He said, "The movie had streamlined the plot and made it so much more effective and made connections that I had never thought to make."

The character of Xena originally appeared on the TV series *Hercules: The Legendary Journeys.* She was so popular, she got her own spin-off series, *Xena: Warrior Princess.*

During the filming of aerial scenes in Greenland for *Dr. Strangelove or: How I Learned to Stopped Worrying and Love the Bomb,* the crew accidentally filmed over a secret U.S. military base. Their plane was forced to land and they were initially thought to be Soviet spies.

- ABBA was originally slated to do the music for Disney's *The Lion King* but couldn't because of a scheduling conflict. Elton John was brought in instead.

- Jackie Chan provided the singing and speaking voice for the Beast in the Chinese dub of *Beauty and the Beast.*

- Rapunzel and Flynn from Disney's *Tangled* can be seen in the background in *Frozen*.

- *Frozen* is the first animated Disney feature to be directed by a woman.

- Joss Whedon cowrote *Toy Story*. The character of Rex was one of his contributions to the script.

- *Pirates of the Caribbean: On Stranger Tides* currently holds the record for the most expensive film ever made with a budget of $378.5 million.

- The most expensive film series ever made is the Hobbit trilogy. The total budget of all 3 films was $623 million.

- P. L. Travers hated the Disney adaptation of her book *Mary Poppins*. She wept during the film's premiere and refused to let Disney adapt her subsequent works.

- *Charlie and the Great Glass Elevator,* the sequel to Roald Dahl's *Charlie and the Chocolate Factory,* was never adapted into a movie because Dahl hated the adaptation of his first book.

The first televised ad for the Got Milk? campaign was directed by Michael Bay.

Barry Manilow's 1975 hit "I Write the Songs" was written by Bruce Johnston.

Mary Poppins was the first film to feature animatronics. Wires in Mary Poppins's dress were used to animate the motion of the birds that sing along with her.

Sleeping Beauty was a total flop when it came out. It cost the Walt Disney Company millions of dollars—this led to massive layoffs and Disney's first year posting a net loss.

Walt Disney's Oscar for *Snow White and the Seven Dwarfs* came with 1 large Oscar statue and 7 miniature ones.

The Great Ormond Street Hospital owns the rights to the Peter Pan books and collects royalties on them. Peter Pan's creator, J. M. Barrie, gave the hospital the rights to his work to ensure it had a stable income.

Who Framed Roger Rabbit, the 1988 family film, is based off the 1981 book *Who Censored Roger Rabbit?* The book is considerably darker, dealing with murder, greed, and racism.

Steven Spielberg is the world's highest-grossing director. As of 2015, his films have generated over $9.2 billion in total world box office sales. Peter Jackson is second with a world box office gross of over $6.5 billion.

Marc Okrand created the language Klingon or Klingonese for use in the *Star Trek* franchise. What started as just a few utterances in the *Star Trek* films and TV series has since been developed into a full language with a dictionary and grammatical rules.

Danny Elfman composed *The Simpsons'* famous theme music in just 2 days.

In the original opening credits for *The Simpsons,* when Maggie is scanned at the grocery store, the price that appears on the cash register is $847.63. This was the estimated cost of raising a baby for 1 month in 1989.

Maggie Simpson is voiced by Elizabeth Taylor in the only episode in which she speaks. She says the word "daddy" in the episode "Lisa's First Word."

The first scripted, interracial kiss on U.S. television occurred on November 22, 1968, during the *Star Trek* episode "Plato's Stepchildren."

Marilyn Manson's real name is Brian Hugh Warner.

Jennifer Yuh Nelson, the director of *Kung Fu Panda 2,* is the world's highest-grossing female director. Her films have made $645 million and counting as of 2015.

William Goldman's 1969 western *Butch Cassidy and the Sundance Kid* was originally titled *The Sundance Kid and Butch Cassidy.* The names got reversed after superstar Paul Newman signed on for the role of Butch.

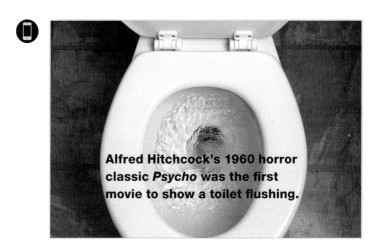

Alfred Hitchcock's 1960 horror classic *Psycho* was the first movie to show a toilet flushing.

- *Showgirls* currently holds the record for the most Razzies nominations ("awards" given for bad films) with 13. Paul Verhoeven was the first director to ever attend a Razzies ceremony and collect his Worst Director award in person.

- *The Dark Knight* made more in its first week than *Batman Begins* made in its entire U.S. run.

- Rock band Pink Floyd performed under many names, including Tea Set, Sigma 6, the Screaming Abdabs, and Leonard's Lodgers, before deciding on Pink Floyd.

- Prince played over 25 different instruments on his debut album, *For You*.

- James Lilja, former drummer of the pop-punk band the Offspring, left the band in 1987 to pursue his dream of becoming a doctor. Today he is a successful gynecologist in California.

- Andre Romelle Young, better known as Dr. Dre, was a diver on his high school's swim team.

- Johnny Marr of the Smiths tried out to play professional soccer for Manchester City FC. He claims he was good enough to play professionally, but the team didn't call him back after his tryout because he was "the only player wearing eyeliner."

Selfie queen Kim Kardashian may look flawless all the time, but the reality star revealed that she suffers from psoriasis, an autoimmune disorder that irritates the skin. Her mother, Kris Jenner, has also dealt with the same condition.

 Drew Carey made a name for himself with self-deprecating fat jokes while playing a chubby midwesterner on *The Drew Carey Show,* but he was a fit and handsome marine before starting his stand-up career.

 Scientists named an albino cave-dwelling daddy longlegs after the *Lord of the Rings* character Sméagol (also known as Gollum).

 Selma Blair played a bus driver on *The Adventures of Pete & Pete*.

Mary-Kate Olsen decorated her wedding ceremony with bowls of cigarettes.

When Tony Sirico landed his role in *Goodfellas,* he'd already been arrested 28 times.

The runner-up of the "Mr. Ugly" contest in Zimbabwe said that he deserved to win because his ugliness was more natural than the winner's.

The programming language Python is named after Monty Python. Developers often make hidden references to Monty Python in their code.

A British family had to be dragged out of their house by firefighters because they refused to leave their burning house. They were too busy watching TV and couldn't wait to get back inside to finish the show when the fire was extinguished.

There was, briefly, a fifth Beatle. Jimmie Nicol replaced Ringo Starr for 7 shows in 1964 when Starr fell ill with tonsillitis. He'd hoped the brief stint would launch his music career, but he never found much success in the industry.

There was 1 real bullet in the gun used in the Russian roulette scenes in *The Deer Hunter,* per Robert De Niro's request.

According to Dan Aykroyd, the original *Blues Brothers* movie with John Belushi had a budget for cocaine.

In Italy, the director of *Cannibal Holocaust* was imprisoned 10 days after the film's release because a local magistrate was convinced that an impalement scene in the movie must have been real. The director then had to track down and fly in the actor who was impaled in order to prove that he had not produced an actual snuff film.

The Department of Veterans Affairs established a special 1-800 hotline for World War II veterans who were traumatized after seeing *Saving Private Ryan*.

Sean Connery turned down the part of Gandalf in the *Lord of the Rings* films—even though he was promised 15% of the box office—because he did not understand the script. That 15% would have earned him $450 million.

Director Eli Roth screened *Cannibal Holocaust* for a lost tribe in the Peruvian Amazon to show them what a movie was. The tribe found the violent horror film hilarious.

Contrary to its portrayal in *Jurassic Park,* the *Tyrannosaurus rex* probably didn't roar. Instead, scientists believe it either hissed or rattled, like a rattlesnake.

Because Han Solo says, "I'll see you in Hell," in *The Empire Strikes Back,* there is a complex mythology for what Hell is in the *Star Wars* universe.

The Godzilla costume from the 1954 film was so thick and heavy that the actor playing Godzilla could be in it for only 3 minutes at a time, and a special valve had to be installed in the suit to drain the actor's sweat.

When the Bill Murray film *Groundhog Day* was released in 1993, a critic at the *Washington Post* wrote derisively that the movie would never be designated a national film treasure by the Library of Congress, but in 2006, it was.

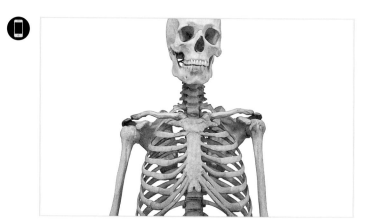

The skeletons in the film *Poltergeist* are real. When asked why, the director explained that they were cheaper than plastic skeletons.

Real-life Roman gladiators frequently performed product endorsements before battles. The makers of the film *Gladiator* originally planned to show this, but the idea was shelved because the producers thought audiences would find the historical fact unbelievable.

Heath Ledger asked Christian Bale to actually beat him up during the interrogation scene in *The Dark Knight*.

During the first season of *The Fresh Prince of Bel-Air,* showrunner Will Smith memorized the lines of everyone in the cast. If you re-watch, you can sometimes catch him lip-synching other characters' lines.

While watching the 1980 film *The Blue Lagoon,* a herpetologist observed an undocumented species of iguana on screen. He then tracked down the species in Fiji and named it the crested iguana.

Joseph Stalin allowed the 1940 film adaptation of *The Grapes of Wrath* to be screened in the Soviet Union because it depicted the oppression of the poor under capitalism. However, it was later banned because Soviet audiences were shocked that even the impoverished Okies in the film had cars.

The screenplay for *Atuk* has never been filmed because every person who had been selected to star in the film died unexpectedly before production began, including John Belushi, John Candy, and Sam Kinison.

When *Ghostbusters* was being shot in New York City, Bill Murray carried around thousands of dollars in cash, which he constantly handed out to the homeless.

Because he portrayed James Bond, Daniel Craig can walk into any Aston Martin showroom and take whatever car he wants for the rest of his life.

A video editor recut every scene in *Lost* chronologically. The entire recut series is available for torrent online.

In the Vietnam protest rally scene in *Forrest Gump* where Tom Hanks's mic is cut, what he actually says is "Sometimes when people go to Vietnam, they go home to their mommas without any legs. Sometimes they don't go home at all. That's a bad thing. That's all I have to say about that."

Pulp Fiction, Forrest Gump, The Lion King, The Shawshank Redemption, and *Jurassic Park* were all in theaters in October 1994.

The original sound track for *2001: A Space Odyssey* was written by composer Alex North. Stanley Kubrick was unhappy with it and changed it at the last minute without telling North, who showed up at the film's premiere and was devastated that not a single second of his work was used.

The sound of Godzilla's roar was made by rubbing a glove over the strings of a double bass.

That "just born" look in movies is created by rubbing babies with a mix of jam and cream cheese.

Eminem was offered the lead role in *Elysium,* but turned it down because the filmmakers refused to shoot it in Detroit, per the rapper's ultimatum.

Throughout the entire Twilight series, there is a total of 24 minutes of the actors just staring.

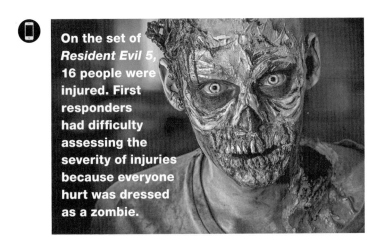

On the set of *Resident Evil 5,* 16 people were injured. First responders had difficulty assessing the severity of injuries because everyone hurt was dressed as a zombie.

Grand Theft Auto V cost $265 million to make—that's more than any Hollywood movie, with the exception of the $378.5 million *Pirates of the Caribbean: On Stranger Tides*.

Dan O'Bannon, the writer of *Alien,* based the scene in which an alien bursts out of a man's chest on the pains he suffered from Crohn's disease.

The song "Circus Galop" was written for player pianos and cannot be played by a human being.

- The 2002 historical drama *Russian Ark* is 1 hour and 36 minutes long and was shot in a single uninterrupted take.

- On the final day of shooting *Titanic,* someone spiked the crew's lobster chowder with PCP, and 80 people were sent to the hospital.

- Dexter Holland, lead singer of the Offspring, is on his way to earning a Ph.D. in molecular biology. He put a hold on his studies when the band got its break, but decided to go back to school after turning 40. His research focuses on HIV.

- Nicolas Cage bought an octopus as a pet. He thought it would help with his acting.

- David Bowie's cocaine problem was so out of hand in 1975 that he claims not to remember recording the critically acclaimed *Station to Station* record.

- In 1996, an armed man broke into a New Zealand radio station and demanded they play "Rainbow Connection" by Kermit the Frog.

- DMX, Notorious B.I.G., Busta Rhymes, and Jay Z all attended George Westinghouse Career and Technical Education High School in Brooklyn, New York.

- In 1972, Pink Floyd recorded a live concert at an amphitheater in Pompeii for a film, but there were no fans in attendance.

The welcome sign in Kurt Cobain's hometown of Aberdeen, Washington, says COME AS YOU ARE.

Many Avril Lavigne fans in Brazil believe that the pop-punk princess died at some point between the release of her first and second albums and was replaced by a doppelgänger named Melissa, who has been acting as Lavigne for the past 10 years.

Chuck Norris's given name is Carlos Ray Norris.

Paul McCartney painted HEY JUDE on a window to promote the Beatles' single of the same name, but members of the local Jewish community mistook it for Nazi graffiti and smashed the window.

During the Continuation War, the Finnish army played a popular polka song to drain the batteries of radio-powered Soviet mines.

Tom Petty, John Mellencamp, and Sting all asked George W. Bush not to use their songs during his 2000 presidential campaign.

Wolfgang Amadeus Mozart wrote a piece called "Leck mich im Arsch," or "lick me in the arse."

Christopher Walken was a circus lion tamer as a teenager. He applied for the job because he "liked cats."

When the U.S. Navy refused filming permits to the director of *Crimson Tide,* the crew waited outside of a naval base until a submarine went out to sea, and then chased it to get a shot of it submerging.

Willy Wonka & the Chocolate Factory was funded by Quaker Oats to support its real-life Wonka Bar, which was taken off shelves after only a few months because of poor sales.

The Jack Black–helmed film *Nacho Libre* was based on a real Catholic priest in Mexico who turned to wrestling in order to raise money for an orphanage, an idea he got from a 1963 Mexican film.

Disney executives initially hated Johnny Depp's performance as Captain Jack Sparrow in *Pirates of the Caribbean* and thought he would ruin the film.

The stars of the *Super Mario Bros.* film, Bob Hoskins and John Leguizamo, drank heavily throughout the shooting because they thought it was so bad.

James Cameron reportedly carried around a nail gun during the filming of *Avatar* and would use it to nail cell phones to the wall if he heard one ring.

Steven Spielberg refused payment for directing *Schindler's List*.

Though she's only 4 feet, 7 inches tall, sex expert Dr. Ruth Westheimer served as a sniper in the Israeli army and was seriously wounded in the 1948 Arab-Israeli War.

Early in his career, Elvis Presley was told by the manager of the Grand Ole Opry that he should stick to truck driving.

A Guns N' Roses show was once delayed because Axl Rose was watching the second Teenage Mutant Ninja Turtles film, *The Secret of the Ooze*.

Wes Craven got the idea for *A Nightmare on Elm Street* after reading about a mysterious pandemic that swept through Southeast Asia in the 1970s, in which otherwise healthy young men would suddenly die in their sleep; it was believed to be related to post-traumatic stress.

Oliver Stone based the final scene in *Platoon* off a real battle he fought in during the Vietnam War.

Miley Cyrus's name at birth was Destiny Hope. Miley is a nickname, short for "Smiley," given to her because she smiled so much as a baby. She legally changed her name in 2008.

Bill Broyles, the writer of *Cast Away,* isolated himself on a desert island for 1 week in order to find inspiration for the film's script. A volleyball washed up onshore at one point, which was the inspiration for the "character" Wilson.

- Jeremy Renner worked as a makeup artist before making it big in Hollywood.

- Haley Joel Osment does not blink on-screen in *A.I. Artificial Intelligence,* because Stephen Spielberg felt a robot would not.

- Seann William Scott, most famous for playing Stifler in the American Pie films, did not have a girlfriend until he was 30.

- Walt Disney was fired from his cartoonist job at a Kansas newspaper because they thought he was uncreative.

- When Queen Elizabeth visited the set of *Game of Thrones,* she refused to sit on the iron throne because she is not allowed to sit on foreign thrones.

- After Samuel L. Jackson signed on to *Snakes on a Plane,* the producers attempted to change the film's name, but Jackson said the only reason he wanted to be in the film was because of the name.

- The first 3-D film was released in 1922, but it is assumed to have been lost or destroyed.

- The idea for *The Terminator* came to James Cameron in a dream during a bad bout of food poisoning.

The FBI sent a team to investigate the production of *Borat* due to multiple reports of a suspicious Middle Eastern man driving an ice-cream truck in the Midwest.

Steven Spielberg refused to direct the Harry Potter films because it wouldn't be a challenge.

Pierce Brosnan worked in a circus as a fire-eater as a teen. He started taking classes because there were a lot of topless women in the class.

In order to fulfill the funerary wishes of rock star Gram Parsons, after his fatal overdose in Joshua Tree National Park, his tour manager and friends stole Parsons's body from the airport and drove it back to Joshua Tree in a stolen hearse. Upon reaching the Cap Rock section of the park, they attempted to cremate Parsons's corpse by pouring 5 gallons of gasoline into the open coffin and throwing a lit match inside.

Brazilian record collector Zero Freitas has a private stock of over 6 million records—the largest in the world.

A Turkish man was arrested for posting a meme on Facebook comparing Turkey's president to Gollum. Charged with "insulting a public official," his trial was put on hold after the judge and prosecutor admitted they had not seen the *Lord of the Rings* movies and couldn't really judge if being compared to Gollum was an insult.

- Truman Capote insulted Robert Frost so deeply that Frost found out where he worked (the *New Yorker* magazine) and had him fired. Capote's offense? Leaving one of Frost's poetry readings early because he had a cold.

- English singer Morrissey wrote an erotic fiction novel, *List of the Lost*.

- Vin Diesel is an anagram for "I End Lives." His legal name is Mark Sinclair.

- Though he starred in a movie called *Chocolat,* Johnny Depp was allergic to chocolate as a child.

- Lou Bega's "Mambo No. 5" was initially chosen for 2000's Democratic National Convention, but this idea was scrapped due to the line "A little bit of Monica in my life."

Nicolas Cage was once stalked by a mime, whom he described as "silent—but maybe deadly."

Rolling Stones guitarist Keith Richards claims he heard the riff to "(I Can't Get No) Satisfaction" in a dream, recorded it, and then went back to sleep.

The U.S. Navy has been using the music of Britney Spears to frighten Somali pirates.

The death metal band Hatebeak's lead singer is an African gray parrot.

Madonna was fired from Dunkin' Donuts for squirting jelly donuts at customers.

5% of people have no emotional reaction to music.

The band Queen never won a Grammy.

In 1986, EMI budgeted the Red Hot Chili Peppers $5,000 to make a demo tape. They did it with $3,000 and spent the rest on hard drugs.

Before launching his daytime television career, Jerry Springer was the mayor of Cincinnati.

As a student protester in the civil rights era, Samuel L. Jackson was part of a group of students that held Martin Luther King Sr. hostage.

Creed once played a show that was so horrible that their fans sued them for $2 million.

Psychedelic musician Jim Sullivan released 1 album in 1969 called *U.F.O.*, then disappeared 6 years later without a trace, never to be heard from again.

Swedish man Roger Tullgren receives disability benefits due to his addiction to heavy metal.

Steve Buscemi was a firefighter before he became a famous actor. He was a member of the FDNY for 4 years and returned to his old company in the days after 9/11 to help search for victims. The FDNY appointed him an honorary battalion chief in 2014.

In 1994, the English acid house band the KLF burned £1 million worth of sterling, the majority of their earnings, in a boathouse on the Scottish island of Jura. While initially unrepentant, KLF member Bill Drummond told the BBC in 2003, "It's a hard one to explain to your kids and it doesn't get any easier. I wish I could explain why I did it so people would understand."

Paul Newman taught Jake Gyllenhaal how to drive.

Kobe Bryant speaks fluent Italian. Bryant's father was also a professional basketball player, though much less successful. After several years in the NBA, Bryant's father moved the family to Italy to continue playing basketball.

By her own account, Nicki Minaj was fired from at least 5 different Red Lobster restaurants. Once she chased after a customer who left with her pen after signing his receipt.

Three Russian women were sentenced to 15 days in jail for twerking in public.

Justin Timberlake's mother was briefly Ryan Gosling's legal guardian.

Americans cast 63 million votes for Taylor Hicks, the winning contestant of *American Idol*'s fifth season finale. Compare that with the 54.5 million votes that won Ronald Reagan the U.S. presidency in 1984.

California man Juan Catalan spent 5 months in jail on a murder charge before he was released and awarded $320,000 in a police misconduct lawsuit. The reason he was cleared? Footage taken at a Dodgers game for Larry David's HBO show, *Curb Your Enthusiasm,* showed Catalan at the stadium, giving him an alibi for the crime.

Inside the Actors Studio's James Lipton had a brief career as a pimp in a Paris bordello in the 1950s.

When Oasis was recording "Fade In-Out" on their album *Be Here Now,* guitarist Noel Gallagher was too high and drunk to play the slide guitar part. Johnny Depp happened to be in the studio at the time and stepped in to play the part.

Though famous for his family-friendly roles, Tim Allen spent 2 years in jail when he was found in possession of 650 grams of cocaine.

Woody Harrelson's father was a hit man. Convicted of murder twice, including the assassination of a federal judge, Charles Harrelson once claimed credit for the assassination of John F. Kennedy (authorities found his cocaine-fueled claims dubious).

Harry Potter author J. K. Rowling was fired from her secretarial job for daydreaming too much.

Dennis Rodman has said that he has 47 siblings.

Gossip Girl star Leighton Meester was born in prison. Her mother was serving a drug sentence at the time for her role in a marijuana smuggling ring in Jamaica.

1. **At the moment John Lennon was pronounced dead, what song was playing over the hospital's sound system?**

   a. "Imagine"

   b. "All My Loving"

   c. "Hold My Hand"

   d. "Hey Jude"

2. **Mary-Kate Olsen decorated her wedding ceremony with bowls of what?**

   a. Apples

   b. Rose stems

   c. Cigarrettes

   d. Candy corn

3. **Throughout the entire *Twilight* series, there is a total of how many minutes of staring between the actors?**

   a. 24

   b. 45

   c. 10

   d. 60

4. **What job did Christopher Walken have as a teenager?**

   a. Seal trainer

   b. Veterinary assistant

   c. Lion tamer

   d. Crab fisher

Food, Animals, Objects, & More!

We can thank bats for tequila. Bats are the biggest pollinators of the agave plant, which is what tequila is made from.

The smallest living dog is a Chihuahua named Miracle Milly, who weighs less than a pound and could fit into a teaspoon when she was born.

Norway knighted a penguin. Named Nils Olav, he is an honorary colonel in chief of the Norwegian King's Guard.

Burrito means "little donkey" in Spanish.

Pigs don't actually sweat; they don't have any sweat glands. In fact, the reason they wallow around in mud is to cool off.

Dogs prefer to defecate when their bodies are aligned along north-south magnetic lines, but don't seem to care where they urinate.

Pigeons have microscopic balls of iron in their ears, which might explain how they are able to navigate.

Days in the winter actually get longer for beavers. Beavers mostly stay in their dams during winter, and because they don't see very much sunlight, their internal clocks shift into 29-hour cycles.

Blue whale calves gain as much as 200 pounds a day.

Government officials in Louisville, Kentucky, sent out an e-mail to department staffers warning of disciplinary action after finding mass quantities of boogers on the walls next to the urinals in the men's bathroom.

There are only 3 white rhinos left in the entire world. They live in a preserve in Kenya and are protected by armed guards.

A zoo in England stopped feeding its monkeys bananas for health reasons.

Eating yogurt or milk along with a hot dog reduces the risk of cancer.

Americans eat nearly 23 sticks of butter each year, although in the 1920s that figure was closer to 72 sticks.

Humans have been making wine for over 9,000 years. But the first wines were usually made with spices, herbs, and even pine resin and probably tasted more like turpentine than merlot.

Olives are teeming with bacteria. Olives that are fermented have the same kind of probiotics that yogurt does.

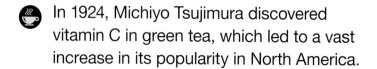

In 1924, Michiyo Tsujimura discovered vitamin C in green tea, which led to a vast increase in its popularity in North America.

One serving of kale has 1180% of the daily recommended value of vitamin K. Vitamin K is named after the German word *koagulation,* which means "blood clotting," because it plays an important role in the regulation of blood clots.

Wild boars wash their food off before eating it, usually with water, but sometimes with spit if they're lazy.

A Wisconsin man sent 240 donuts to his local police department after cops kicked him out of a football game. The police donated the donuts to the Salvation Army, citing a healthy eating initiative. Also, the donuts were coconut flavored, which none of the cops in question liked.

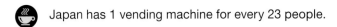 Japan has 1 vending machine for every 23 people.

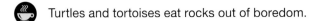 Turtles and tortoises eat rocks out of boredom.

The microwave was invented by accident when engineer Percy Spencer walked by a radar set and the candy bar in his pocket melted. The first food cooked by a microwave on purpose was popcorn.

40% of all food in America ends up in landfills instead of stomachs. That averages out to about 20 pounds per person every month.

Due to a genetic defect, cats can't taste sweet things.

There are 11 miles of underground tunnels beneath Los Angeles. During Prohibition, many of them became speakeasies.

The original owner of the Red Lobster chain first got into the restaurant business with a lunch counter called the Green Frog.

Eight classic Corvettes were swallowed by a giant sinkhole that opened up under the National Corvette Museum.

A minister and his wife sued a blind man and his seeing-eye dog after the dog allegedly stepped on her toes and broke them. According to witnesses, the woman made no move to get out of the dog's way because she wanted to see if the dog would walk around her.

You can break in a new pair of leather shoes by aiming a hair dryer at them.

 A new monkey species was discovered in Bolivia in 2004. The scientists who discovered it auctioned off the naming rights to support the national park it lives in. The winner was an online casino, and the monkey is known as the GoldenPalace.com monkey.

 The price of Coca-Cola remained at 5¢ for over 70 years. A big reason for this was that Coke vending machines were built to only accept nickels. At one point the head of the company lobbied President Eisenhower to make 7.5¢ coins because he thought increasing the price to a dime would be too much of a shock.

 An Ohio woman unhappy with her Chinese food order called 911 to complain. Instead of getting her money back, she was arrested.

 A Peruvian bulldog named Otto holds the world record for skateboarding through the longest human tunnel.

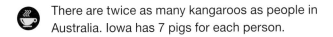 There are twice as many kangaroos as people in Australia. Iowa has 7 pigs for each person.

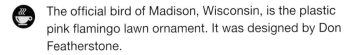 The official bird of Madison, Wisconsin, is the plastic pink flamingo lawn ornament. It was designed by Don Featherstone.

The U.S. Postal Service employs donkeys to deliver mail to the bottom of the Grand Canyon.

Lemmings do not actually follow each other off a cliff to their deaths. The myth comes from a Disney "documentary" produced in 1958 in which the filmmakers threw the lemmings off a cliff themselves.

Nashville has a coin-operated used-car vending machine.

Salt-rising bread smells like feet—instead of yeast, it uses the same bacteria that cause gangrene. Its origins are murky, but it was probably developed by pioneers in the Appalachians in the beginning of the 19th century.

There is a popular Middle Eastern conspiracy theory accusing Israel of using sharks with electronic devices attached to them as weapons against Palestine.

In 2015, 50% of the population of Saiga antelopes died en masse, and nobody is certain why.

The 4-millimeter-long pseudo-scorpion inhabits the pages of old books, protecting them by eating the microscopic bugs that would otherwise feed on the paper.

Dolphins have the longest memory among nonhuman animals and can recognize the whistles of familiar dolphins after having been separated for 20 years.

At 188 decibels, the blue whale produces the loudest sound of any animal on Earth.

The possum can neutralize almost any poison, and scientists are currently working to apply its ability to treat humans suffering from poisoning.

The disclaimer "no animals were harmed during the making of this film" can be applied to the credits of any production, even if animals were killed, as long as they didn't die while the cameras were actually filming.

It's true that the vampire bat sucks the blood of its victims. It's also true, though less commonly known, that the vampire bat urinates on its victims while sucking their blood.

The saber-tooth squirrel in the 2002 animated children's movie *Ice Age* was thought to be fictional. However, the fossils of a real one were discovered in 2011.

The Japanese macaque monkey makes and throws snowballs for entertainment.

The Chinese giant salamander can grow up to 6 feet long and emits an odor similar to black pepper.

If a goose becomes ill or injured during flight, 2 other geese will leave the formation to protect their friend until it recovers or dies.

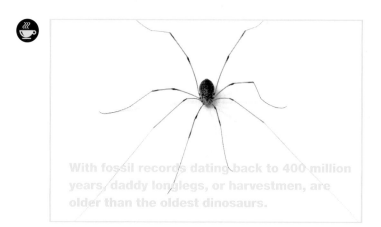

With fossil records dating back to 400 million years, daddy longlegs, or harvestmen, are older than the oldest dinosaurs.

Crickets have twice the amount of protein ounce for ounce than beef and 3 times the amount of iron.

Kraft Macaroni and Cheese is more popular in Canada than anywhere else in the world.

The "Bambi effect" is a term used anecdotally or in editorial media that refers to objections against the killing of animals perceived as cute, such as deer or dolphins.

Movie theater popcorn costs more per ounce than filet mignon.

Nestlé has sold over 200 flavors of Kit Kat bars in Japan, including soy sauce, crème brûlée, green tea, and banana.

Regular Coke sinks in water because of the heavy sugar, while Diet Coke floats.

A man named Cooter Brown stayed drunk during the entire Civil War to avoid being drafted.

Alcohol is prohibited in the British Parliament with 1 exception: the chancellor can drink while delivering the annual budget statement.

On average, every person older than 15 in colonial America drank the equivalent of 7 shots of alcohol per day.

A bonobo named Kanzi once asked for "fire" and "marshmallows" using lexigrams while on a hike in the woods. His trainer gave him matches and some marshmallows. He broke twigs, made a fire, and roasted the marshmallows over it.

A Japanese zoo unsuccessfully tried to mate a pair of hyenas for 4 years before realizing both were males.

Koalas are the only animal—besides humans and great apes—that have unique fingerprints.

The word "sommelier" used to refer to a butler in charge of the provisions and storeroom of a large estate. In today's lexicon it simply means a wine expert.

Many historians believe the custom of clinking glasses and saying "Cheers!" or "To health!" originated in the Middle Ages. Because poisoning one's foes was so common in the Middle Ages, people would often take some wine from their cup and pour it into the cup of whomever they were sharing a drink with, ensuring that if one glass had been poisoned, both would be. The glasses would often clink together during this process. As time went on and people became less concerned with poisoning, the custom of pouring wine from one glass into the other died away, but the tradition of clinking glasses together and reciting a salutation stuck around.

The earliest known instance of winemaking was discovered in the Zagros mountain region of modern Iran. Archaeologists there found pottery that had been used to store and age wine dating back to 5400 B.C.

Americans consume the most wine on Thanksgiving Day, more than any other day of the year.

The oldest known bottle of wine was discovered in the tomb of a Roman noble in Speyer, Germany, in 1867. It is thought to date from around 325 B.C. It is currently on display at Historisches Museum der Pfalz in Germany.

Vatican City has the highest per capita wine consumption in the entire world, according to the Wine Institute.

A glass of wine has roughly 1 cluster, or 75 grapes, in it. A bottle of wine has 4 clusters, which comes out to about 2.5 pounds of grapes.

Champagne has to come from the Champagne region of France, or else it must be labeled as sparkling wine.

The Napa Valley draws more tourists to California each year than Disneyland.

Brotherhood Winery in Washingtonville, New York, is acknowledged as the oldest continually operating winery in the United States. It produced its first vintage in 1839. Brotherhood survived the Prohibition era by making sacramental wine and nonalcoholic beverages.

Ancient Greeks found it uncouth and uncivilized to drink wine straight. They served their wine mixed with equal parts water, often fortified with herbs.

The term "blanc de noir" literally translates to "white of black" or "white of darkness." It refers to white wine made from red or black grapes.

Wine bottles comes in 13 standard sizes:

> Piccolo—.1875 liters
> Demi—.375 liters
> Standard—.750 liters
> Magnum—1.5 liters
> Jeroboam or Double Magnum—3 liters
> Rehoboam—4.5 liters
> Methuselah or Imperial—6 liters
> Salmanazar—9 liters
> Balthazar—12 liters
> Nebuchadnezzar—15 liters
> Melchior or Solomon—18 liters
> Primat or Goliath—27 liters
> Melchizedek or Midas—30 liters

 Incitatus—the favorite horse of Roman emperor Gaius, most commonly known as Caligula—was given his own seat in the Roman senate.

 Black cats were first publicly denounced and associated with devil worship by Pope Gregory IX. The resulting slaughter of cats has been blamed for the bubonic plague, as there were fewer cats to kill off the rats, which host the plague-carrying fleas.

 All worker bees are female and so is the queen. Drone bees are male and don't have a stinger.

Bees have 2 stomachs: 1 for eating and 1 for storing nectar.

 The average honeybee will make 1/12 of a teaspoon of honey in its lifetime.

 Bees use their honeycomb cells to raise their offspring in and to store nectar, honey, pollen, and water.

 The chocolate chip cookie was invented by Ruth Graves Wakefield in the mid-1930s. She owned the Toll House Inn, a popular restaurant in Whitman, Massachusetts. Wakefield claims to have invented the chocolate chip cookie while riffing on a classic butterscotch nut cookie recipe. To this day, every bag of Nestlé chocolate chips sold in North America has her original recipe on it.

 A group of hippopotamuses is called a bloat or thunder.

 A group of squirrels is called a dray or scurry.

A Wisconsin man aged a block of cheddar for 40 years before selling it. It was said to be quite sharp.

A group of bears is called a sleuth or sloth.

A group of moles is called a labour.

The expiration date on water is not the date the water will go bad. It is the estimated date the chemicals in the bottle might leach into the water and affect the taste.

Jellyfish are 95% water.

Giraffes can go longer without drinking water than camels.

The McDonald's in Sedona, Arizona, is the only one in the world with turquoise arches instead of golden ones. The turquoise coloration was thought to be a better fit for the reddish desert surroundings.

There is a basketball court in the Supreme Court building. It is often called the "highest court in the land" because it sits directly above the courtroom.

Fred Baur, inventor of the Pringles potato chip package, had his ashes interred in a Pringles can after he died.

The highest bill still printed is the $100 note. However, in the past there were $500, $1,000, $5,000, and even $10,000 notes. Few of these large-denomination notes remain in circulation. As of 2009, the Federal Reserve estimates only 336 $10,000 bills were known to exist, 342 $5,000 bills, and 165,372 $1,000 bills. They are prized collector's items and often sell for much higher than their face value. William McKinley is on the $500 bill, Grover Cleveland is on the $1,000, James Madison is on the $5,000, and Salmon P. Chase is on the $10,000. Although no large-denomination bills have been printed since 1969, they are still legal tender and you could use one today.

 3M researchers Spencer Silver and Art Fry are credited with inventing the Post-it Note. Both men had worked to create a weak adhesive but couldn't find a use for it, until one day, while at his chorus rehearsal, Fry realized his musical notes kept falling out and had the idea for the Post-it Note.

 The potato chip was invented by Chef George Crum. One night in 1853, Crum had a difficult customer who kept sending back his fried potatoes, claiming they were not crispy enough. Crum eventually sliced the potatoes into paper-thin sheets and deep-fried them. They were a hit with the fussy customer and with everyone else who tried them, and the potato chip was born.

The serving of ice cream in a waffle cone dates back to the 1904 St. Louis World's Fair. It was an unusually hot day, which meant ice-cream vendors were doing a brisk business while hot waffle vendors couldn't find interested customers. At one point an ice-cream vendor ran out of cups to serve his ice cream in. A waffle vendor then rolled up one of his waffles and suggested he serve his ice cream out of that. The collaboration was a hit.

Pound cake gets its name from the original recipe, which called for a pound of each of the following ingredients: butter, sugar, eggs, and flour.

Dunkin' Donuts locations in South Korea offer kimchi croquettes and glazed garlic-flavored donuts.

The tea bag was originally invented not for the preparation of tea, but for the delivery and distribution of tea samples.

Improperly prepared fugu, or puffer fish, can kill you. The fish contains a potent toxin—1,200 times stronger than cyanide—that must be removed by a skilled chef prior to serving. It is considered a delicacy in Japan.

McDonald's sells 75 hamburgers every second of every day on average.

One fast-food hamburger can contain meat from 100 cows or more.

Peanuts are not technically a nut; they are a legume.

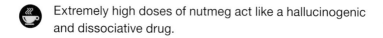 Extremely high doses of nutmeg act like a hallucinogenic and dissociative drug.

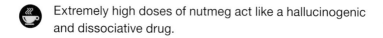 Yams and sweet potatoes are not the same thing; they are two separate and distinct root vegetables.

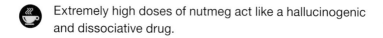 Cucumbers are about 96% water.

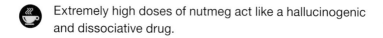 At birth, baby kangaroos are only about 1 inch long, smaller than many insects.

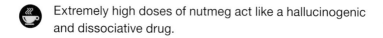 A single elephant tooth can weigh up to 9 pounds.

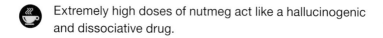 Both sheep and rabbits are immune to the bites of black widow spiders.

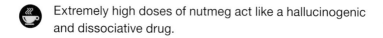 Sharks kill roughly 10 people a year while humans kill millions of sharks a year.

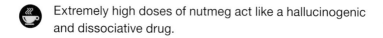 Dogs have a sense of smell that is about 100,000 times more powerful than that of a human.

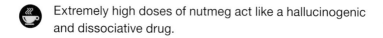 Cats are capable of mind control, but not in an evil scientist sort of way. Research has found that cats can change the pitch of their meows to sound more like crying babies and manipulate us into giving them food, attention, and so on.

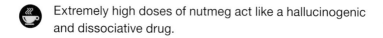 Polar bears have blackish skin under their white fur.

Gentoo penguins propose to their potential mates with supersmooth pebbles.

Fish develop a fever when they get emotional.

Domesticated cats typically hate water, but wild leopards are quite graceful swimmers.

Pet mini-pigs have been growing in popularity recently, but are often abandoned when the pigs become too large. Some unscrupulous breeders pass off pigs originally intended for agriculture as pet pig breeds.

A convenience store clerk in Florida thwarted a would-be robber dressed in a Darth Vader costume when he threw a bottle of bleu cheese salad dressing at Darth's head.

The record for solving a Rubik's Cube the fastest, just 4.9 seconds, is held by fourteen-year-old Lucas Etter.

Agricultural scientists have bred a spiceless jalapeño pepper for reasons unknown.

A Taco Bell in California offers valet parking.

The world's rarest primate is the Hainan gibbon. Prized for their purported medicinal properties, the gibbons were often boiled whole to make a tonic paste. There are only 28 left, and they live on a small island off China's southern coast.

The smartest horse in the world is a former racehorse named Lukas who can solve simple math and spelling problems.

A village in Ghana is home to a group of extraordinarily docile crocodiles. Locals believe the crocodiles hold the souls of their ancestors. The beasts have never harmed anyone and even let people pet them.

The platypus isn't the only mammal that lays eggs. Echidnas, which look like anteaters but are not related, also lay eggs. Males have 4-headed penises. They are distantly related to platypuses, but evolutionarily diverged 19–48 million years ago.

Most of Los Angeles's palm trees are not native to California. Many of the city's iconic trees were planted during the Great Depression as part of a public works program.

- Pebble toads get their name from their tendency to curl up in a ball when threatened, and they often roll down hills.

- Orange juice requires analyzing up to 1 quintillion decision variables in order to deliver a consistent product that always tastes the same, making it one of the most complex business operations in the world.

- There have been at least 7 recorded instances of people being struck by flying cows. Fortunately, most of them were not fatal (to the people at least).

- A Frosted Flake in the shape of Illinois sold on eBay for $1,350.

- Pepsi started a riot in the Philippines in 1993 when a lottery offering 1 million pesos (about $40,000) mistakenly had 800,000 winners. Armed men started throwing Molotov cocktails at Pepsi plants, and the incident ended up costing Pepsi millions.

- The most expensive coffee in the world is made from beans that are found in elephant poop. The beans sell for as much as $1,100 a kilo.

- The first Kentucky Fried Chicken was based in Salt Lake City, Utah, and Colonel Sanders was originally from Indiana. Though Sanders's first restaurant was based in Kentucky, it was called Sanders Court and Café.

- It is possible to extract vanilla flavoring from cow poop.

The International Organization for Standardization, Technical Committee 34 (food products), Subcommittee 8 (tea) published the BS 6008:1980, also known as ISO 3103, which details the proper British way to brew a cup of tea. It created a small controversy for failing to mention the pre-warming of the teapot.

Hakarl is an Icelandic delicacy made of fermented shark meat buried underground for up to 12 weeks. Although technically edible, it contains high amounts of ammonia. It's usually chased with a shot of schnapps.

Mosquitos are equally attracted to the smells of Limburger cheese and human feet.

Full beer bottles break more easily than empty beer bottles.

Cows that are named tend to produce more milk.

Dimes have 118 ridges on their edges while quarters have 119.

The official color of the Golden Gate Bridge is called International Orange.

Burgers served in many public school lunches have 26 ingredients.

Chimpanzees can identify other chimps by looking at their butts.

Durian is a large spiky fruit native to Southeast Asia that some people find delicious but most people think smells like turpentine or raw sewage. The smell is so overpowering that the fruit has been banned on many Asian public transportation systems.

Buddha Jumps Over the Wall is a fabled shark fin soup said to be so delicious that vegetarian monks would hop over monastery walls to get it.

"Wonder grain" quinoa is not actually a grain. Technically it's a chenopod and is related to beets, spinach, and tumbleweeds.

It's common for decaf coffeepots to have orange handles because of branding efforts by the instant decaf coffee brand Sanka. Sanka derives its name from the French term for "decaffeinated," *sans caféine*.

"Taser" and "laser" are both acronyms. TASER stands for Thomas A. Swift's Electric Rifle. LASER stands for Light Amplification by Stimulated Emission of Radiation.

In 2012, the newly released Flamin' Hot Cheetos caused a health scare. Parents brought their children to emergency rooms in droves thinking they had blood in their stool due to the Flamin' Hot Cheetos' food coloring.

- The 2 pilots on any British Airways flight are not allowed to order the same meal, lest it should be contaminated with bacteria and put passengers in danger.

- Paul Brown, the inventor of the upside-down ketchup bottle, earned $13 million for his invention when Heinz bought the copyright from him.

- Traditional balsamic vinegar can cost $100 an ounce, because it takes between 12 and 25 years to produce and requires being moved through 7 different barrels of increasingly smaller sizes.

- In South Korea, there is a trend of gastronomic voyeurism. One of the most popular faces of the movement is a woman who makes $9,000 per month by eating in front of her webcam.

- Usain Bolt, the Jamaican sprinter who is currently the fastest person in the world, lived on 100 McDonald's chicken nuggets per day during the Beijing Olympics, where he took home 3 gold medals.

- North Koreans love chocolate pies that are manufactured in South Korea. The beloved pies go for about $9.50 on the North Korean black market.

- In 1943, there was a brief ban on presliced bread in the United States as part of the wartime conservation effort. Americans being great appreciators of sliced bread, the ban was widely unpopular and was lifted before the end of the year.

In January 2014, American competitive eater Molly Schuyler broke the world record for consuming a 72-ounce steak in the least amount of time—3 minutes.

Until 1964, chicken wings were considered a waste product and discarded. Teresa Bellissimo, co-owner of the Anchor Bar in Buffalo, New York, had the idea to fry the wings and dip them in her husband Frank's sauce recipe, thus inventing buffalo wings.

Green olives are sold in glass jars, but black olives only come in cans. The reason is that black olives are ripe, whereas green olives are pickled before they ripen. Storing black olives safely requires heating them to 240°F, which glass jars can't sustain. There were mass outbreaks of fatal botulism when people first started selling black olives in jars.

The worst food poisoning outbreak in modern history occurred in 1993, when 73 Jack in the Box restaurants infected 732 people with a particularly virulent strain of *E. coli,* which killed 4 children and left many people with permanent kidney damage.

During the 2011 Norwegian butter crisis, an acute local shortage of the condiment caused prices to inflate to $77 per smuggled stick on the black market.

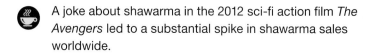 A joke about shawarma in the 2012 sci-fi action film *The Avengers* led to a substantial spike in shawarma sales worldwide.

The ruling in the 1942 U.S. Supreme Court case *Wickard v. Filburn* said the federal government can legally destroy wheat crops grown in excess, even wheat for private consumption.

There is a counterfeit egg industry in China. Apparently an average counterfeiter can produce 1,500 fake eggs per day, which are then sold to unwitting merchants or consumers.

The unique taste of New York City's pizza has been attributed to the flavor of the public tap water supply.

Tomato

KETCHUP

300 ml

France banned ketchup from public schools in 2011, considering the condiment to be a threat to the preservation of the nation's culinary traditions.

A sealed jar of raw honey does not rot. Archaeologists have found 3,000-year-old bottles of the stuff that were still edible.

In 2013, Kim Jong Il's former head chef Kenji Fujimoto revealed that the deceased Great Leader enjoyed feasting on hippos, snakes, and spiders.

When dining in restaurants in Hanoi, Vietnam, you can order the delicacy beverage cobra wine. Your waiter will kill a live cobra and drain its blood into a glass of rice wine, then place the snake's still-beating heart atop the glass for your dining pleasure.

In the 1830s, Greek governor Ioannis Kapodistrias tried to popularize the potato among the people, but to no avail. Kapodistrias then installed armed guards at potato shipment points to make the Greeks think the potatoes were important. It worked, and today the potato is popular throughout Greece.

In the United States, food companies often use carbon monoxide to make meat appear fresher. This practice is illegal in the European Union, Japan, Singapore, and Canada.

Ketchup has been around since about 300 B.C., but tomatoes were not featured in any recipes until 1812, because earlier generations believed the tomato poisonous.

Frito-Lay decided to discontinue its 100% biodegradable SunChips bags because customers complained the crumpled bags were excessively noisy. However, the company unveiled a new design in 2011 that is both biodegradable and less noisy than its predecessor.

The Pilgrims at Plymouth Rock likely dined on swan and seal meat at the first Thanksgiving dinner.

In the 19th century, it was considered cruel and unusual punishment to serve lobster to prisoners, as lobsters were basically the 1800s equivalent of rats.

In 2011, 2 South Florida Domino's Pizza managers were charged with felony arson for burning down a rival Papa John's Pizza franchise in the same town.

Researchers at Newcastle University found that eating a bacon sandwich can cure a hangover. The reaction between amino acids in the bacon and reducing sugars in the fat can alleviate headaches.

Researchers have found that the most arousing food scent to men is pumpkin pie, which can increase penile blood flow by up to 40%.

The Canadian government has a 243,000-square-foot warehouse where it holds a global strategic supply of maple syrup. As of 2015, the warehouse contained 15 million pounds of the sweet and sticky condiment.

According to research funded by Lufthansa, people consume more tomato juice on airplanes than they do on the ground because altitude significantly alters the way they perceive the juice's flavor.

The Golden Temple in India has a kitchen that serves 100,000 free vegetarian meals every day to anyone who asks, regardless of religion, race, or class.

American Airlines saved $40,000 in 1 year by removing 1 olive from each of its in-flight salads.

Because of a 40-year-old holiday marketing campaign, KFC is so popular on Christmas in Japan that customers must order their meals at least 2 months in advance. Christmas isn't even a national holiday in Japan.

Germany consumes more chocolate per capita than any other nation on Earth. The average German citizen consumes 114 chocolate bars per year. To put that in perspective, the average American eats 51 per year.

American explorer William Seabrook dined on human meat with the cannibalistic Guere tribe of West Africa. He reported that it was not like any meat he'd ever consumed, but most closely resembled good veal.

Growth hormones to stimulate milk production in cows are not permitted in Canada under law.

A McDonald's Caesar salad is considerably more fattening than its hamburger, even one served with fries.

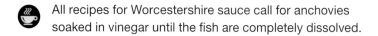 All recipes for Worcestershire sauce call for anchovies soaked in vinegar until the fish are completely dissolved.

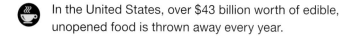 In the United States, over $43 billion worth of edible, unopened food is thrown away every year.

American fruits and vegetables contain significantly less vitamins and minerals than they did 40 years ago due to modern agricultural farming techniques. Iron levels, for example, have declined by 37%.

Taco Bell has tried and failed to enter the Mexican restaurant market on 2 separate occasions. On its second try, the company attempted to market its cuisine to Mexicans as American food.

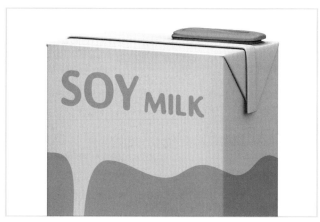

Soy milk does not need to be refrigerated, but grocery stores keep it in the refrigerated foods section to give consumers the impression that it is fresh.

About 4 million cats are eaten every year in China, mostly in the Guangdong and Guangxi Provinces. Cat meat is not considered an acceptable food in northern China.

Ketchup made from mashed bananas and dyed red is very popular in the Philippines. It was developed during World War II when a tomato shortage inspired an enterprising food technologist.

The Breatharian Institute of America, a group that many would call a cult, believes that people can survive on only sunlight and air.

In the United Kingdom, 95% of people polled said the primary reason they purchased organic salads was to avoid consuming pesticides. However, organic food can and typically does contain pesticides.

A farmer in Japan figured out how to grow watermelons in glass cases to create square-shaped versions of the fruit that better fit on store shelves.

A team of German scientists published findings that suggest people's preferences for meat are likely genetically predetermined.

According to the Centre for Retail Research, cheese is the most commonly shoplifted food in the world.

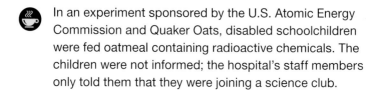

In an experiment sponsored by the U.S. Atomic Energy Commission and Quaker Oats, disabled schoolchildren were fed oatmeal containing radioactive chemicals. The children were not informed; the hospital's staff members only told them that they were joining a science club.

The first ice creams were made in China around 3000 B.C.

New findings by researchers at the University of California at San Diego say that people who eat more chocolate tend to be skinnier on average than those who eat less.

If you eat too much rabbit without other food in your diet, you can actually die, because the body must use a great deal of its own vitamins and minerals to digest the animal's meat.

Moose cheese costs about $420 per pound, because it takes over 2 hours to milk a single moose and the process has to be done in complete silence, lest the animal gets spooked.

In 2006, the *New York Times* reported on xenophobic French soup kitchens that serve only soup made with pork, in order to prevent feeding Muslims or Jews.

Wild salmon are pink because of the crustaceans they eat. In order to match consumer expectations, farmers pick the color they want out of a swatch book called a Salmon Fan. The dyeing process accounts for up to a third of the total cost of raising a salmon.

Gourmand syndrome is a rare, benign condition that can occur when someone injures their right frontal lobe. The victim may develop a new passion for gourmet food—obsessively eating, talking, and thinking about gourmet foods.

Consuming fructose sugar before drinking can increase the speed of alcohol metabolism by 80%.

The first take-out restaurants were street carts in ancient Rome.

The secret spice used in KFC's Original Recipe is sold by Marion-Kay Spices as 99 X.

The name of the office supply chain Staples is a pun. Not only does it sell actual metal staples, but also everyday staples for the home and office.

Cats are asleep for 70% of their lives.

The female kangaroo has 3 vaginas.

A shrimp's heart is located in its head.

Even today, most Chinese citizens consider cheese something eaten by nomadic barbarians living on the country's western borderlands.

American eggs would be illegal in British supermarkets because they are washed. The opposite is also true— British eggs would be illegal in American supermarkets because they are not washed.

Mayor Stubbs has been at the helm of the small town of Talkeetna, Alaska, for 15 years—an impressive feat for any elected official, but even more so considering Stubbs is a cat.

People can theoretically survive on just potatoes and butter.

The *Laetiporus* mushroom tastes almost exactly like chicken.

Artificial raspberry flavoring comes from the anal glands of the beaver.

Asexual chef Mao Sugiyama removed and cooked his genitals, then served them to diners at a special banquet.

A polar bear's liver contains as much vitamin A as you would expect to find in 79–115 chicken eggs. If consumed by a human, the polar bear liver causes hypervitaminosis A, leading to drowsiness, bone pain, peeling of the skin, and, in severe cases, coma or death.

The largest penis on Earth belongs to the blue whale, whose member averages 8–10 feet in length and 12–14 inches in diameter.

The world's oldest creature was a 507-year-old mollusk, which was accidentally killed by scientists.

The oldest soup recipe is 6,000 years old and calls for hippopotamus meat.

Oklahoma's state vegetable is the watermelon.

The Namco Namja Town amusement park in Tokyo offers horse-meat-flavored ice cream.

The Sardinian cheese called *casu marzu* is intentionally infested with maggots, which can leap off the cheese up to 6 inches. The maggots produce enzymes that aid in fermentation. Because the larvae can potentially survive stomach acids and burrow through stomach walls, *casu marzu* is illegal under Italian and European Union law.

The dominant source of flavor enhancer L-cysteine is hog hair.

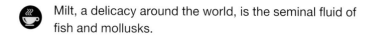 Milt, a delicacy around the world, is the seminal fluid of fish and mollusks.

 The least healthy fast-food fries are Dairy Queen's regular fries, which clock in at 730 calories.

Chuck E. Cheese's pizza restaurants were created by Nolan Bushnell, the mastermind inventor behind the Atari video game system.

Spam was named by the actor Kenneth Daigneau, who was paid $100 for christening the canned mystery meat. According to Spam manufacturer Hormel, the name stands for "spiced ham," though apparently Hormel previously stated that it stood for "shoulder and pork ham."

The apple is a member of the rose family.

Apples float in water because 25% of their volume is air.

The Popsicle was invented by 11-year-old Bay Area resident Frank Epperson in 1905.

The reason peppers taste so hot is because they contain a chemical compound called capsaicin, which "tricks" sensory nerves into thinking they're being burned by fire.

In 1964, a congressional resolution declared bourbon "America's native spirit."

A Canadian woman was arrested on charges of criminal mischief for attempting to give pigs water on a hot day.

M&M'S were originally included in army rations during World War II and not available to the public until after the war.

You can tell if a beer glass is dirty if there are bubbles coming up from the sides of the glass.

Bermuda has an island totally overrun by a group of feral swimming pigs. It's assumed they established themselves on the island after a shipwreck.

Eggo frozen waffles were originally named Froffles. "Leggo my Froffle" just doesn't have the same ring to it . . .

The British Royal Navy gave its sailors a daily ration of rum until 1970. It was served from a special barrel inscribed with the words THE QUEEN, GOD BLESS HER and served midday.

1. **Chimpanzees can identify other chimps by looking at their what?**

    a. Faces

    b. Eyes

    c. Butts

    d. Excretion

2. **The most shoplifted food in the world is what?**

    a. Cheese

    b. Milk

    c. Chocolate

    d. Spinach

3. **A female kangaroo has how many vaginas?**

    a. 1

    b. 2

    c. 3

    d. 4

4. **Oklahoma's state vegetable is a what?**

    a. Broccoli

    b. Carrot

    c. Cabbage

    d. Watermelon

Image Credits

Grateful acknowledgment is made to the following for the use of the photographs that appear throughout the text:

1   eyeball © air009/Shutterstock, Inc.
2   African finch © R. Maximiliane/Shutterstock, Inc.
3   vampire bat © belizar/Shutterstock, Inc.
4   metal ball © 123dartist/Shutterstock, Inc.
5   fingerprint © Andrey_Kuzmin/Shutterstock, Inc.
6   airplane window © 06photo/Shutterstock, Inc.
7   brain © Sergey Nivens/Shutterstock, Inc.
9   hair © Kittibowornphatnon/Shuttersock, Inc.
10   salamander © Steve Bower/Shutterstock, Inc.
12   tarantula © Cathy Keifer/Shutterstock, Inc.
14   sun © Triff/Shutterstock, Inc.
15   DNA © vitstudio/Shutterstock, Inc.
17   dopamine © Ivan Ursul/Shutterstock, Inc.
18   candle © r.classen/Shutterstock, Inc.
19   tequila © Evgeny Karandaev/Shutterstock, Inc.
25   solar system © Vadim Sadovski/Shutterstock, Inc.

28   corpse flower © LiliGraphie/Shutterstock, Inc.

29   dung beetle © paulrommer/Shutterstock, Inc.

30   brain freeze © Linda Bucklin/Shutterstock, Inc.

32   crash dummy © current value/Shutterstock, Inc.

40   pirate flag © sharpner/Shutterstock, Inc.

42   onions © D7INAMTI7S/Shutterstock, Inc.

44   bagpipes © Gorosi/Shutterstock, Inc.

46   King Tut © Merydolla/Shutterstock, Inc.

49   chimp © Everett Collection/Shutterstock, Inc.

52   letters © Preto Perola/Shutterstock, Inc.

56   White House © Vacclav/Shutterstock, Inc.

60   elephant © gualtiero boffi/Shutterstock, Inc.

62   rocks © Valentina Razumova/Shutterstock, Inc.

64   herring © Fanfo/Shutterstock, Inc.

66   Genghis Khan © Andrey Burmakin/Shutterstock, Inc.

67   Brooklyn Bridge © Jenny Lilly/Shutterstock, Inc.

70   Maine sign © Joseph Sohm/Shutterstock, Inc.

75   Van Gogh © vladimir salman/Shutterstock, Inc.

76   piano © glebTv/Shutterstock, Inc.

79   doll © Jakub Krechowicz/Shutterstock, Inc.

81   plane crash © Patrick K. Campbell/Shutterstock, Inc.

84   Pinocchio © Dada Photos/Shutterstock, Inc.

86   lily pond © leisuretime70/Shutterstock, Inc.

90   Vlad the Impaler © Velazquez77/Shutterstock, Inc.

93   The Thinker © Sean Nel/Shutterstock, Inc.

102   cave painting © Eduardo Rivero/Shutterstock, Inc.

105   child napping © CroMary/Shutterstock, Inc.

106   coffee © Valentyn Volkov/Shutterstock, Inc.

111   stamps © paseven/Shutterstock, Inc.

115   hand washing © Walther S/Shutterstock, Inc.

116   honey jar © Thomas Francois/Shutterstock, Inc.

121   soda bottles © Oleksiy Mark/Shutterstock, Inc.

122   nails © Tania Zbrodko/Shutterstock, Inc.

124   mouth © Valentina Razumova/Shutterstock, Inc.

127   tongue © Antonion Guillem/Shutterstock, Inc.

129   beard © SFIO CRACHO/Shutterstock, Inc.

131   swearing © Pranch/Shutterstock, Inc.

133   posture © Undrey/Shutterstock, Inc.

137   feet © Levente Gyori/Shutterstock, Inc.

139   heart © Sebastian Kaulitzki/Shutterstock, Inc.

142   bra © Gulgun Ozaktas/Shutterstock, Inc.

145   yawn © Ollyy/Shutterstock, Inc.

147   crying © Ollyy/Shutterstock, Inc.

151   thumbs-up © x9626/Shutterstock, Inc.

154   pizza © danm12/Shutterstock, Inc.

156   mirror © Evgeny Atamanenko/Shuttestock, Inc.

157   toys © irin-k/Shutterstock, Inc.

158   pilot © g-stockstudio/Shutterstock, Inc.

159   credit card © Pictoores/Shutterstock, Inc.

161   fast food © Mindscape studio/Shutterstock, Inc.

163   robot © Ociacia/Shutterstock, Inc.

164   stadium © Csaba Peterdi/Shutterstock, Inc.

166   pills © Chatkul/Shutterstock, Inc.

168   living room © Zastolskiy Victor/Shutterstock, Inc.

169   hallucinations © ysuel/Shutterstock, Inc.

170   librarian © wavebreakmedia/Shutterstock, Inc.

173   traffic © Artens/Shutterstock, Inc.

174   vending machine © Ibooo7/Shutterstock, Inc.

175   tall vs. short © F. JIMENEZ MECA/Shutterstock

176   smartphone © tanatat/Shutterstock, Inc.

178   sun and earth © Aphelleon/Shutterstock, Inc.

179   walking dog © Rohappy/Shutterstock, Inc.

185   emoji © bontom/Shuttestock, Inc.

187   *Tyrannosaurus rex* © DM7/Shutterstock, Inc.

191   binary computer code © Jason Winter/Shutterstock, Inc.

197   Mary Poppins © zafigachy/Shutterstock, Inc.

199   toilet flush © Feng Yu/Shutterstock, Inc.

201   cigarettes © Givaga/Shutterstock, Inc.

204   skeleton © Potapov Alexander/Shutterstock, Inc.

207   zombie © Kiselez Andrey Valerevich/Shutterstock, Inc.

214   mime © Kiselez Andrey Valerevich/Shutterstock, Inc.

215   jelly donuts © 5/Shutterstock, Inc.

222   pig © jadimages/Shutterstock, Inc.

223   olives © Dionisvera/Shutterstock, Inc.

224   wild boar © Eric Isselee/Shutterstock, Inc.

226   leather shoes © artjazz/Shutterstock, Inc.

227   donkey © Coprid/Shutterstock, Inc.

229   daddy longlegs © Melinda Fawver/Shutterstock, Inc.

234   bee © Peter Waters/Shutterstock, Inc.

240   palm trees © Champ008/Shutterstock, Inc.

246   ketchup © Macrovector/Shutterstock, Inc.

250   soy milk © 3DSguru/Shutterstock, Inc.

253   kitten © Anna Grigorjeva/Shutterstock, Inc.

257   waffles © baibaz/Shutterstock, Inc.

About the Author

When Kris Sanchez joined Twitter in 2009, he used it as a way to connect with Britney Spears. After dropping out of college, Kris aspired to be a dancer, worked with kids, and enjoyed looking up useless facts online. By 2011, Kris had committed himself to tweeting 24/7. By 2012, UberFacts had amassed more than 200,000 followers and Kris was able to secure advertising partnerships that catapulted the social media handle to success.

Today, UberFacts has become a full-fledged brand. @UberFacts boasts more than 13 million Twitter followers and has expanded into a website and a smartphone app. UberFacts has solidified its standing as one of the 150 most followed handles on Twitter and has been supported by Kim Kardashian, Khloé Kardashian, Miley Cyrus, Paris Hilton, Kendall Jenner, Perez Hilton, Ariana Grande, Enrique Iglesias, Aaron Paul, Jack Dorsey, and more.

UberFacts and the man behind it have been featured in *Fast Company, Adweek, Success, Business Insider, Millennial Magazine,* the *Today* show, and the Fox News channel.